SHELTER

SHELTER

by Moki

GINGKO PRESS

Fragile Dwelling

Preface by Margaret Morton

Asleep in doorways, curled atop sidewalk steam vents, stretched out on benches in public parks, perched along rivers and canals, harbored beneath bridges or hidden deep within underground tunnels, the presence of those called "homeless" demands that we confront not only the bleak consequences of economic inequality, but also the humanity of those who strive to create shelters for themselves from the most meager resources. Their assemblages of shipping crates, scrap wood and broken furniture, discards of the modern city, are in fact homes, laboriously and ingeniously built, little by little, piece by piece.

By demand, not by design, private spaces are made public. One can see in these homes – laid bare by circumstance – not only a need for shelter, security, and privacy, but also the profound human need that we all share to personalize, to decorate, to embellish, to collect and to display. Resourceful by necessity, poetic by desire, these improvised dwellings reveal both the dreams and inventiveness of their builders.

Some of these shelters exist for only a few hours; others endure for weeks, months, or even years and evolve into more permanent settlements. Inevitably, the makeshift homes are demolished and the communities formed by these men and women are fractured. Dispossessed, they journey the streets alone, nomads forever on the move: riding subways throughout the night; sleeping on dark, silent streets; hiding in the shadows of construction sites; tucking themselves into decaying structures along the waterfront; disappearing before dawn.

Excerpts from The Poetics of Space

by Gaston Bachelard

It is striking that even in our homes, where there is light, our consciousness of well-being should call for comparison with animals in their shelters. An example may be found in the following lines by the painter, Vlaminck, who, when he wrote them, was living quietly in the country: "The well-being I feel, seated in front of my fire, while bad weather rages out-of-doors, is entirely animal. A rat in its hole, a rabbit in its burrow, cows in the stable, must all feel the same contentment that I feel." Thus, well-being takes us back to the primitiveness of the refuge. Physically, the creature endowed with sense of refuge, huddles up to itself, takes to cover, hides away, lies snug, concealed. If we were to look among the wealth of our vocabulary for verbs that express the dynamics of retreat, we should find images based on animal movements of withdrawal, movements that are engraved in our muscles. p 91

…

A nest, like any other image of rest and quiet, is immediately associated with the image of a simple house. When we pass from the image of a nest to the image of a house, and vice versa, it can only be in an atmosphere of simplicity. Van Gogh who painted numerous nests, as well as numerous peasant cottages, wrote to his brother: "The cottage, with its thatched roof, made me think of a wren's nest." For a painter it is probably twice as interesting if, while painting a nest, he dreams of a cottage and, while painting a cottage, he dreams of a nest. It is as though one dreamed twice, in two registers, when one dreams of an image cluster such as this. For the simplest image is doubled; it is itself and something else than itself. Van Gogh's thatched cottages are overladen with thatch. Thick, coarsely plaited straw emphasizes the will to provide shelter by extending well beyond the walls. Indeed, in this stance, among all the shelter virtues, the roof is the dominant evidence. Under the roof's covering the walls are of earth and stone. The openings are low. A thatched cottage is set on the ground like a nest in a field. p 98

…

If we go deeper into daydreams of nests, we soon encounter a sort of paradox of sensibility. A nest – and this we understand right away – is a precarious thing, and yet it sets us to daydreaming of security. Why does this obvious precariousness not arrest daydreams of this kind? The answer to this paradox is simple: when we dream, we are phenomenologists without realizing it. In a sort of naïve way, we relive the instinct of the bird, taking pleasure in accentuating the mimetic features of the green nest in green leaves. We definitely saw it, but we say that it was well hidden. This center of animal life is concealed by the immense volume of vegetable life. The nest is a lyrical bouquet of leaves. It participates in the peace of the vegetable world. It is a point in the atmosphere of happiness that always surrounds large trees. p 102-103

…

Our house, apprehended in its dream potentiality, becomes a nest in the world, and we shall live there in complete confidence if, in our dreams, we really participate in the sense of security of our first home. In order to experience this confidence, which is deeply graven in our sleep, there is no need to enumerate material reasons for confidence. The nest, quite as much as the oneiric house, and the oneiric house quite as much as the nest – if we ourselves are at the origin of our dreams – knows nothing of hostility of the world. Human life starts with refreshing sleep, and all the eggs in the nest are kept nicely warm. The experience of the hostility of the world – and consequently, our dreams of defence and aggressiveness – come much later. p 103-104

…

I have simply wanted to show that whenever life seeks to shelter, protect, cover or hide itself, the imagination sympathizes with the being that inhabits the protected space. The imagination experiences protection in all its nuances of security, from life in the most material of shells, to more subtle concealment through imitation of surfaces. As the poet Noël Arnaud expresses it, being seeks dissimulation in similarity. To be in safety under cover of a color is carrying the tranquility of inhabiting to the point of culmination, not to say, imprudence. Shade too, can be inhabited. p 132

Refugee

by OI VA VOI

You ask me
Why it is I come to you
When someone else is just as good
I asked them but they said the same
Didn't even ask my name

Explain to me
Just what it is you have to lose
Take a minute in my shoes
Don't you feel like you've paid your dues
Already

I'll show you
That all our fates are so entwined
Don't lose your faith in humankind
Just don't forget my state of mind
Is fragile

Together
We can enjoy the taste of dignity
As long as you believe in me
I'll show you my reality
I've seen a few

There's another refugee
There's another refugee
Oh oh there's another refugee
Oh oh there's another refugee

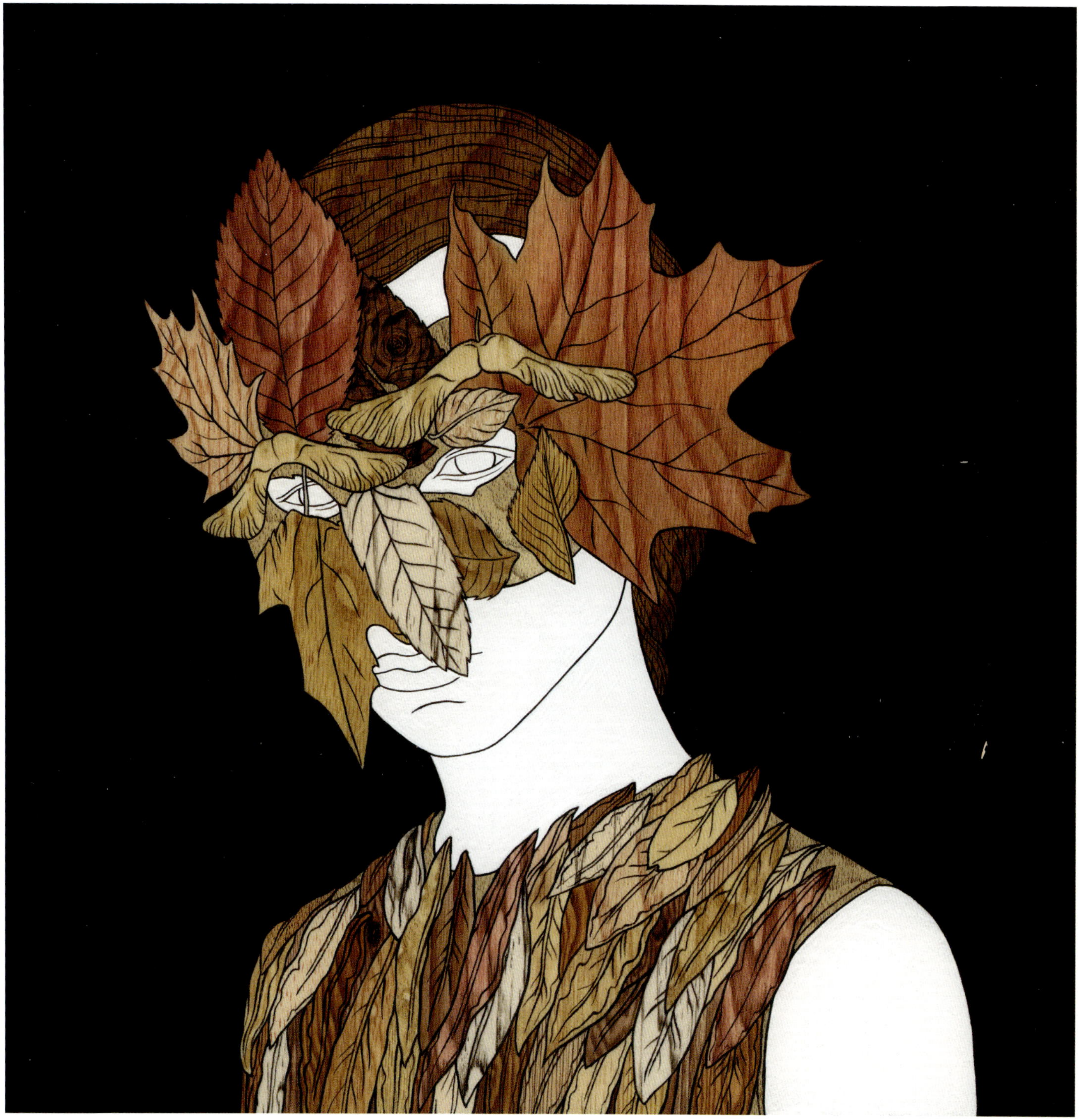

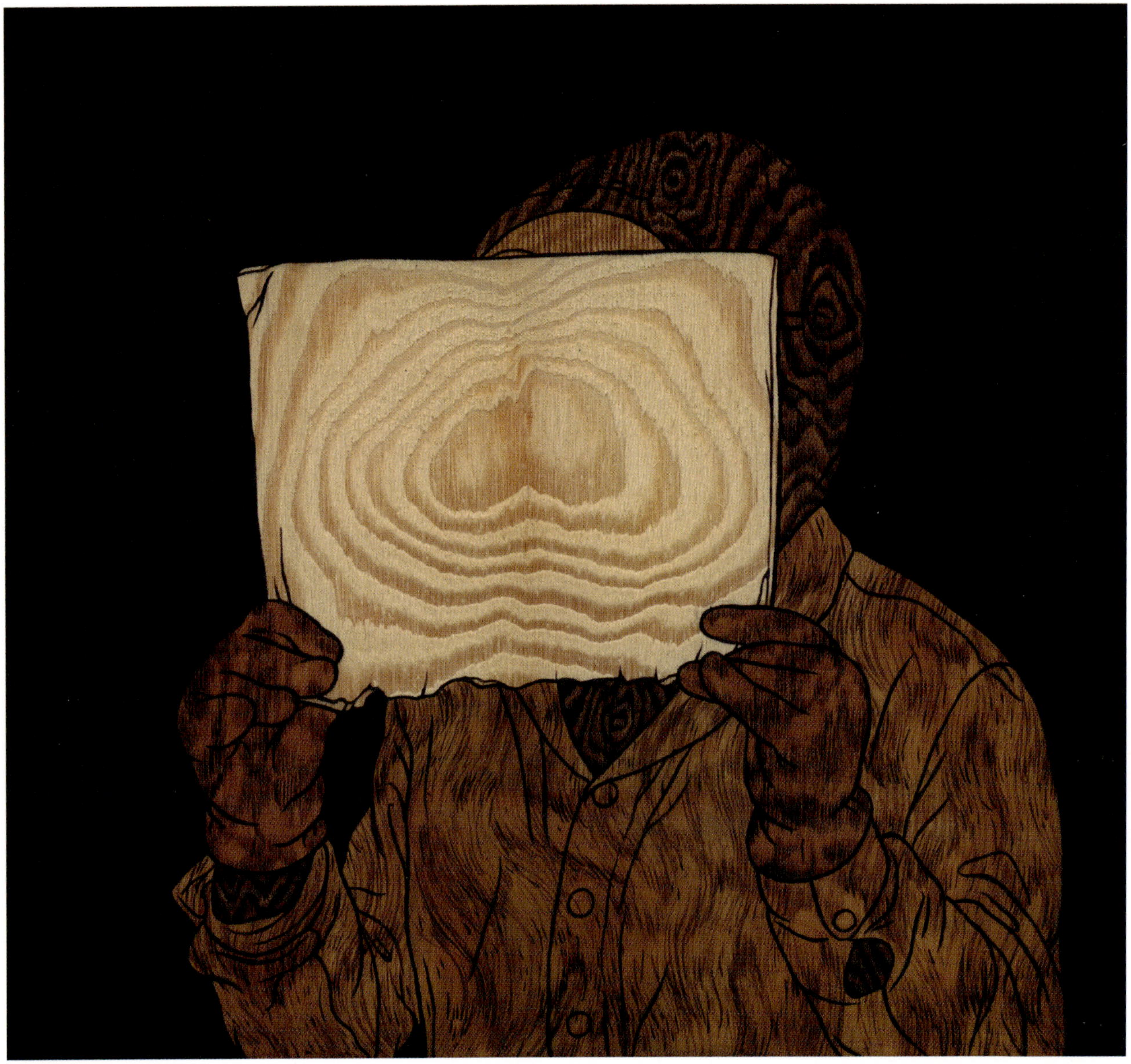

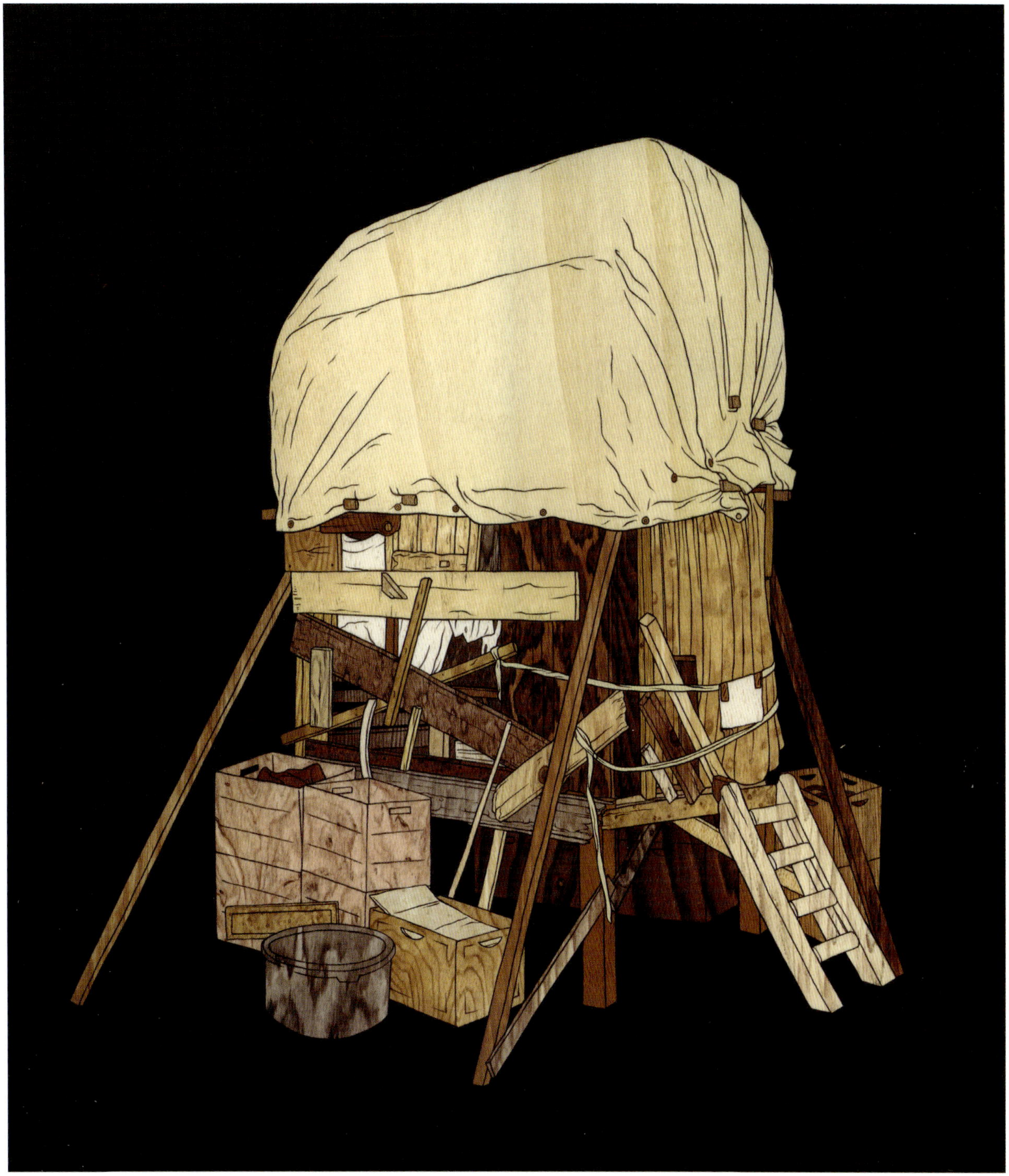

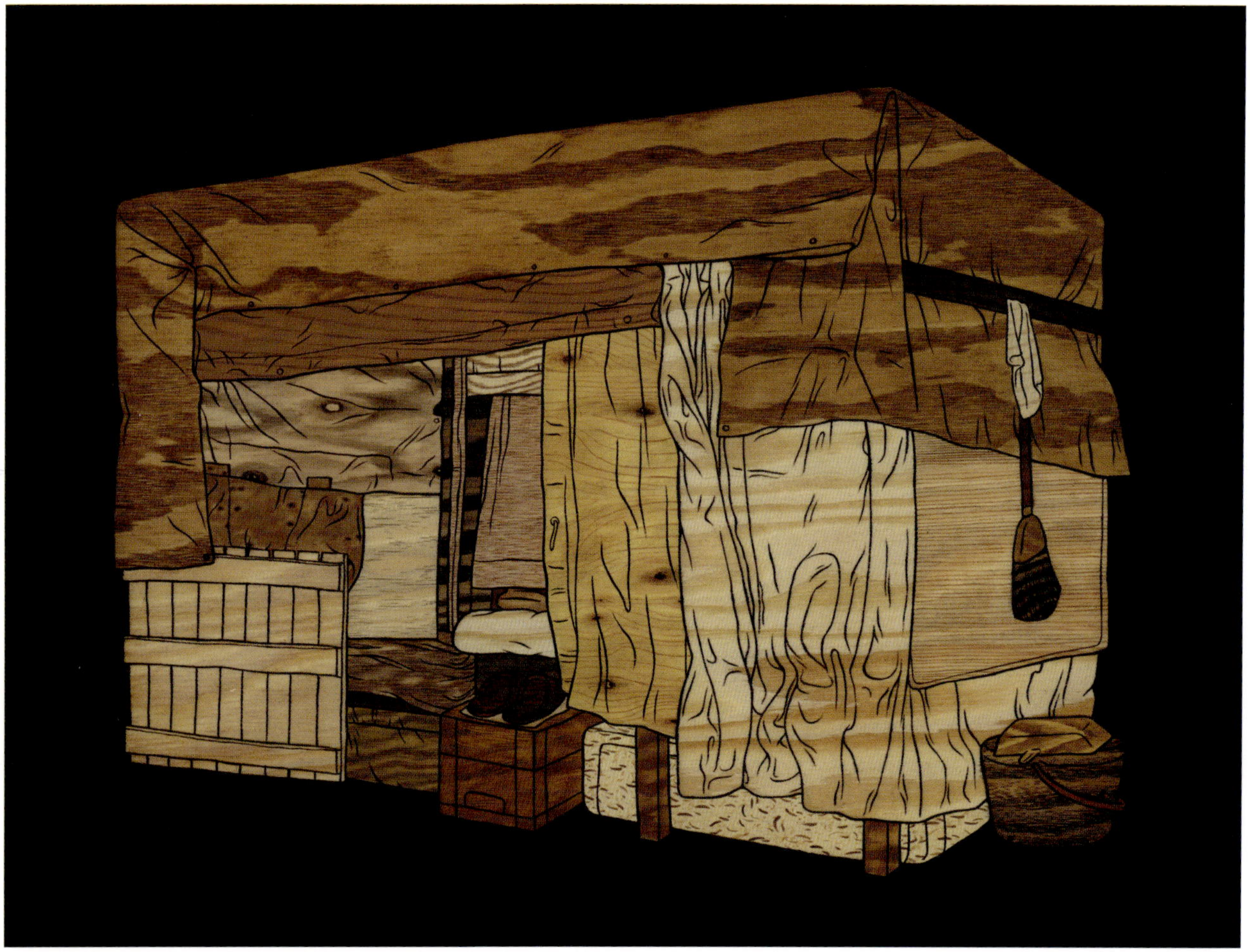

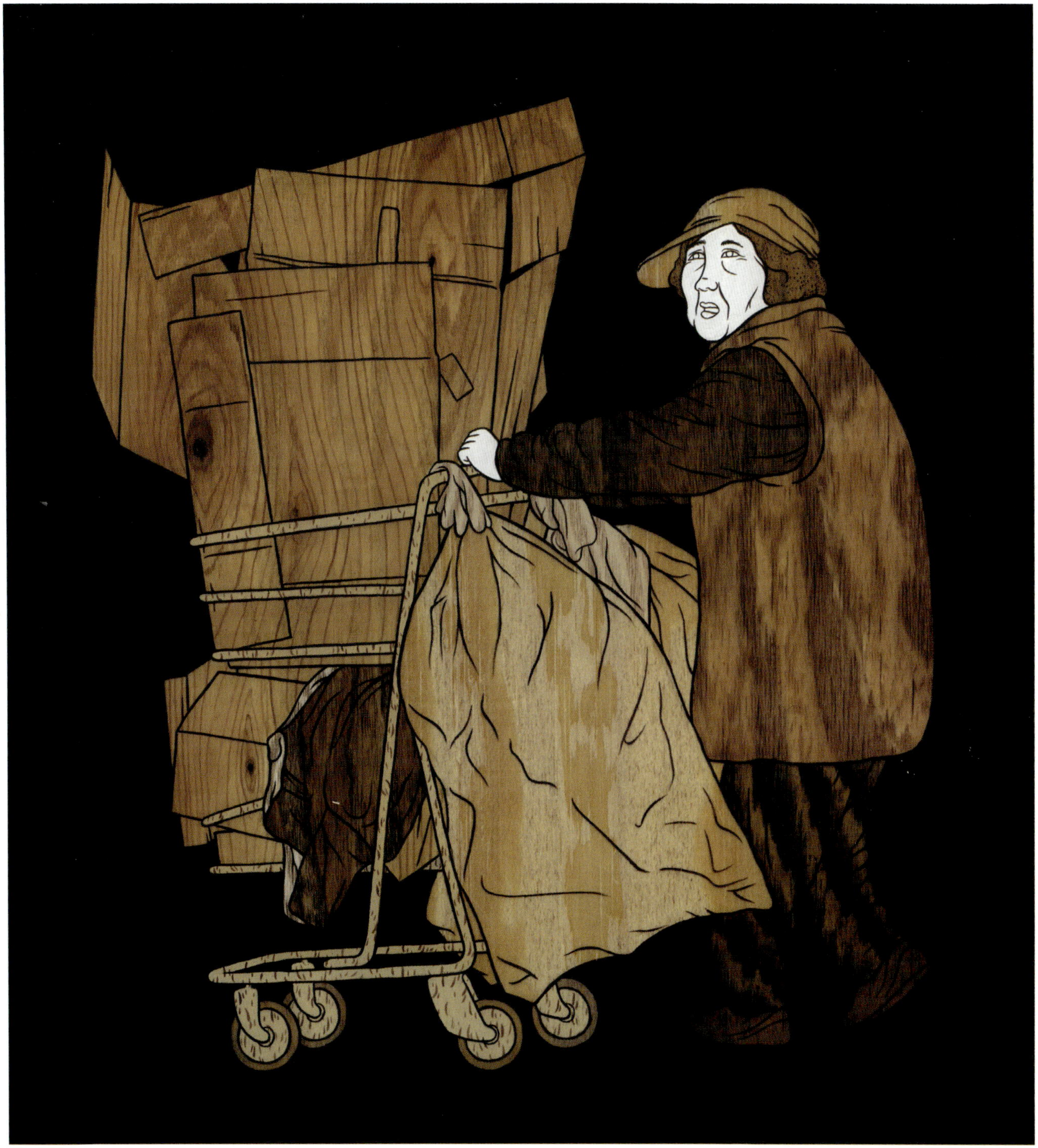

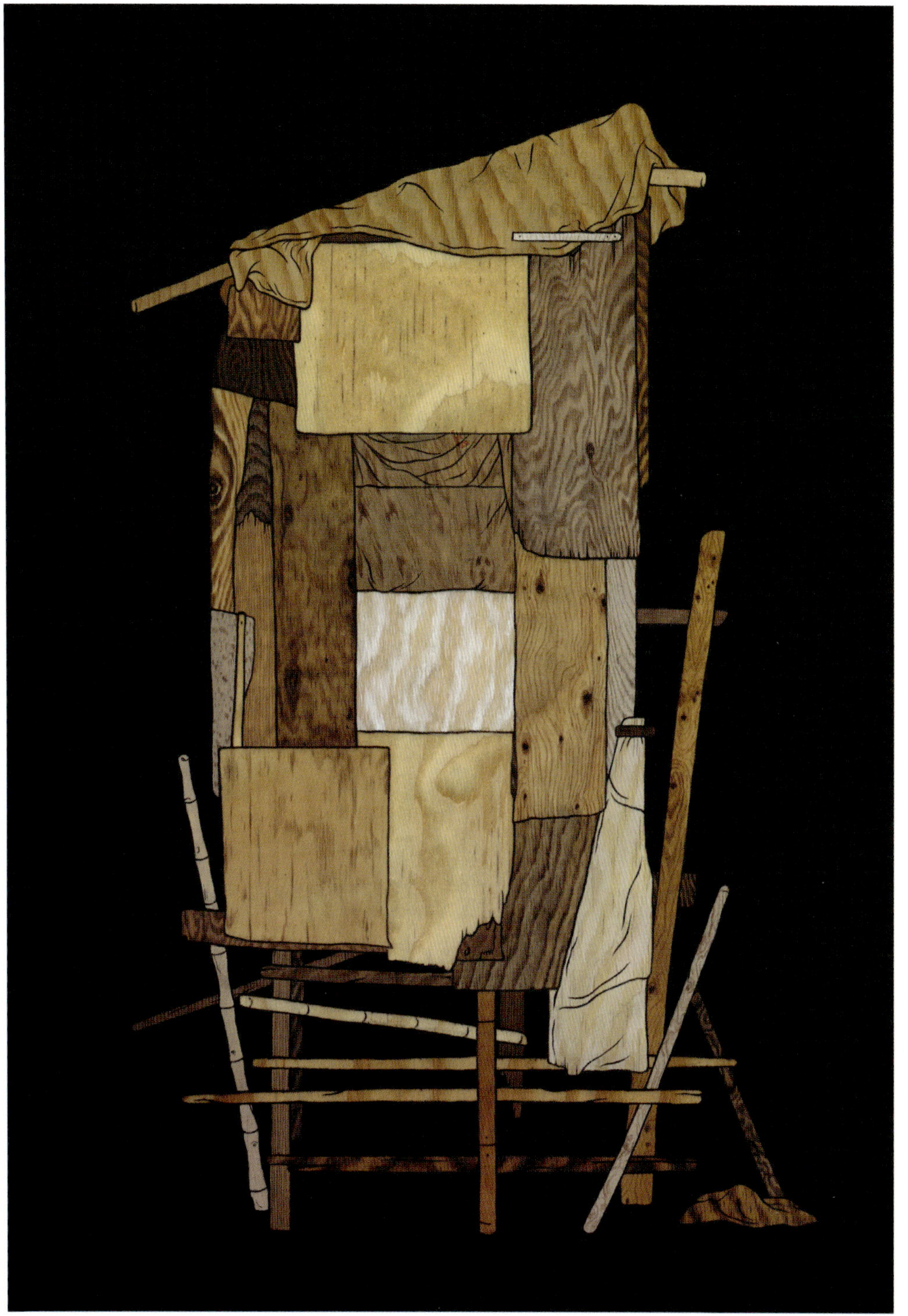

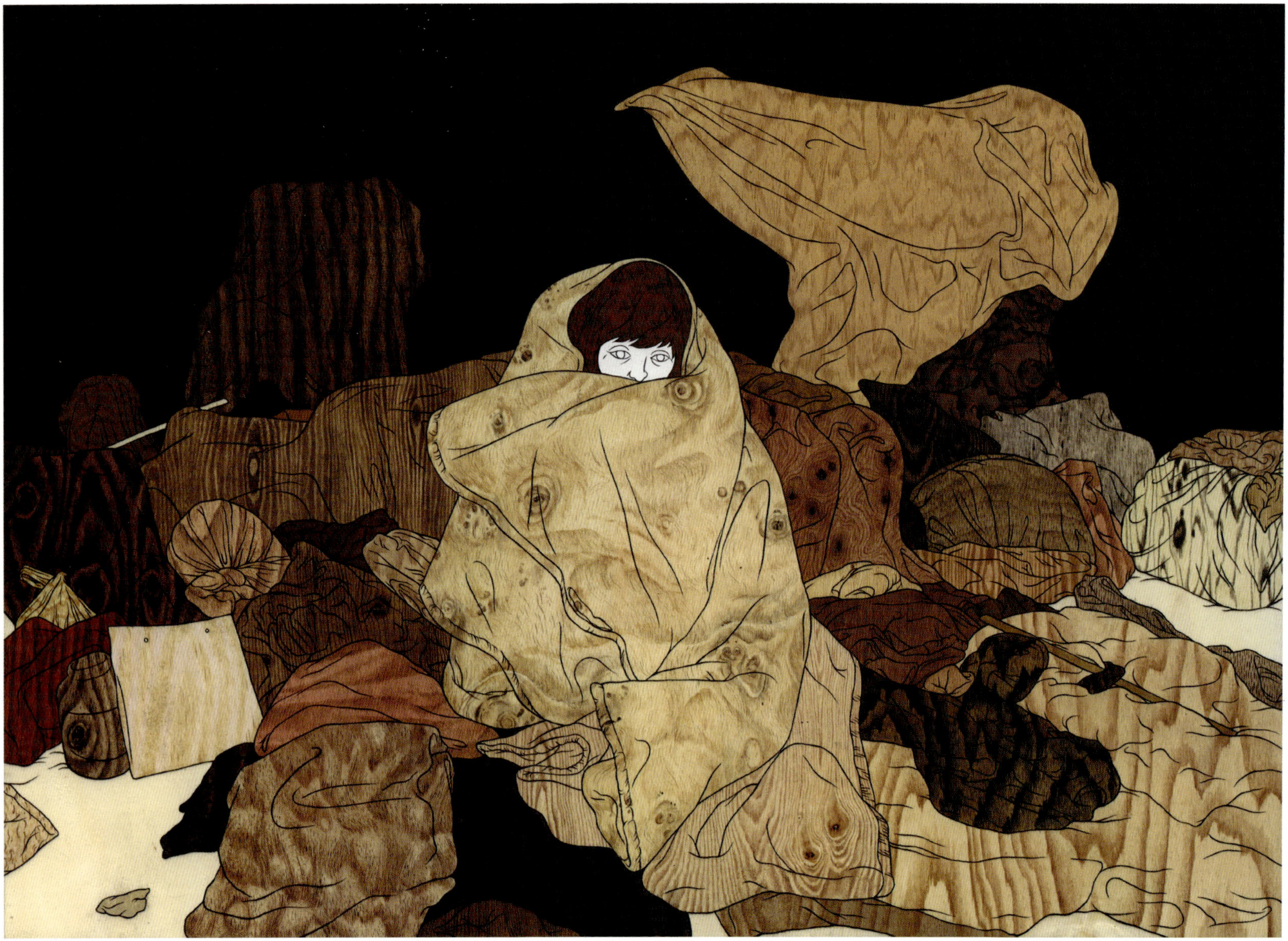

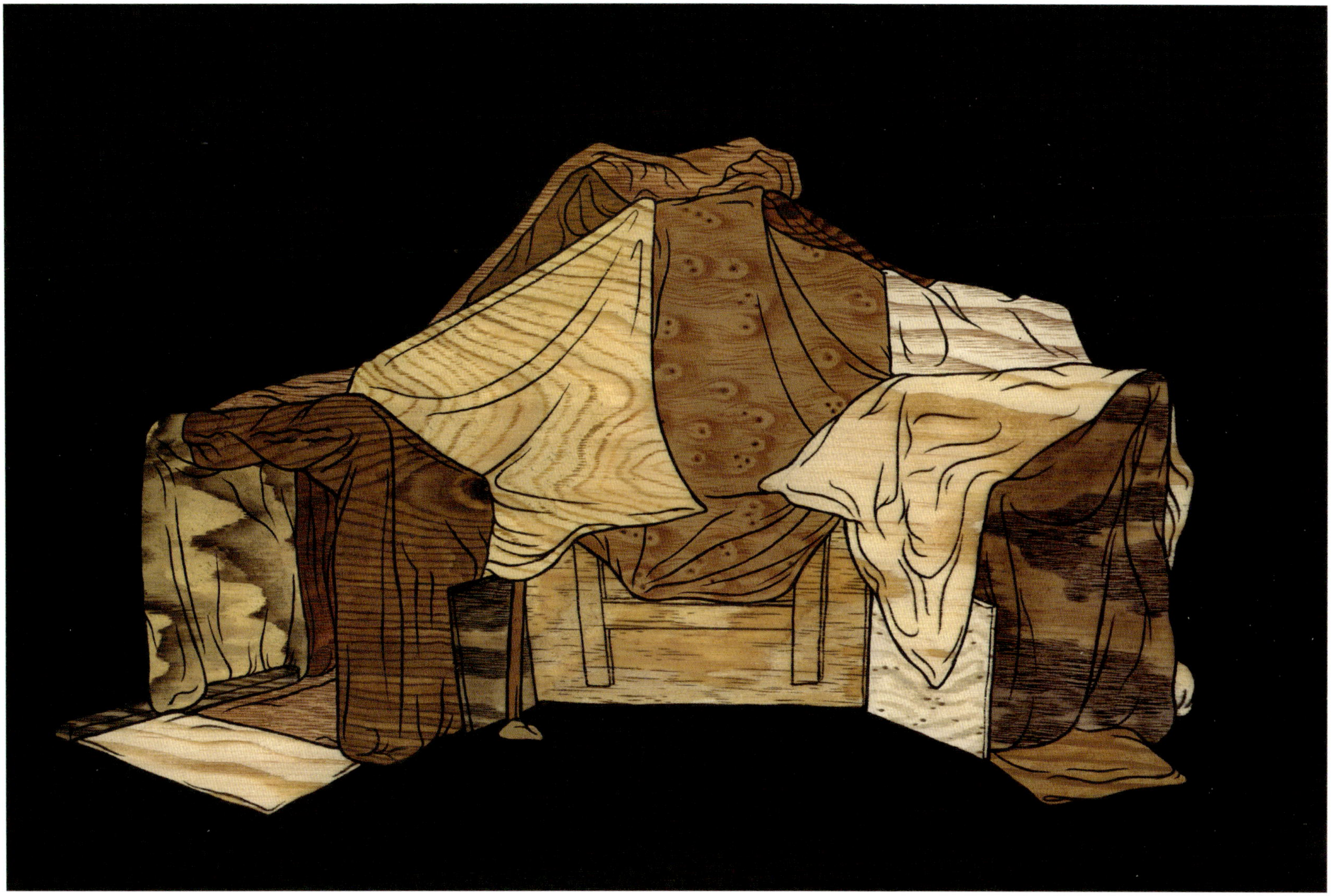

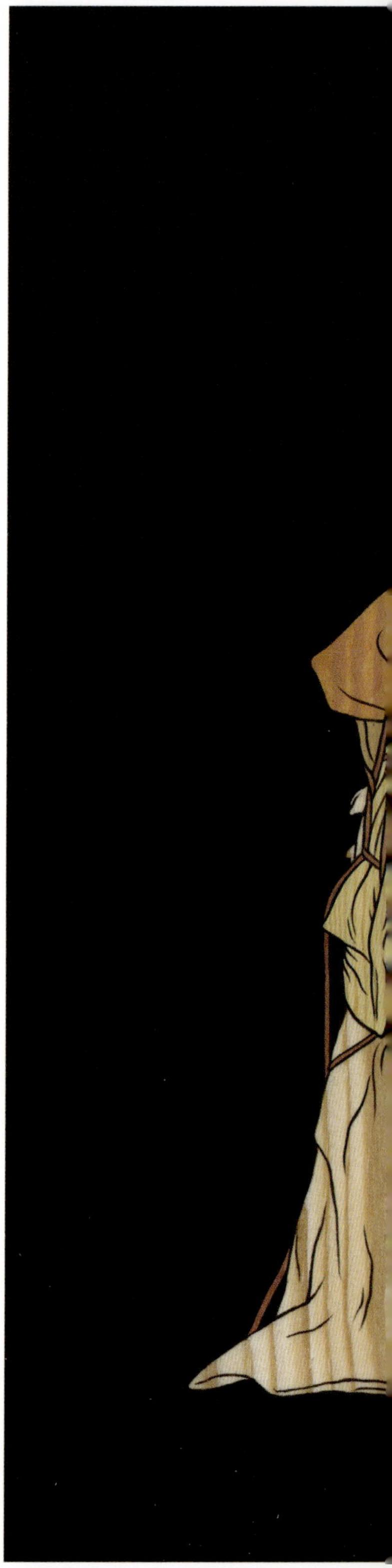

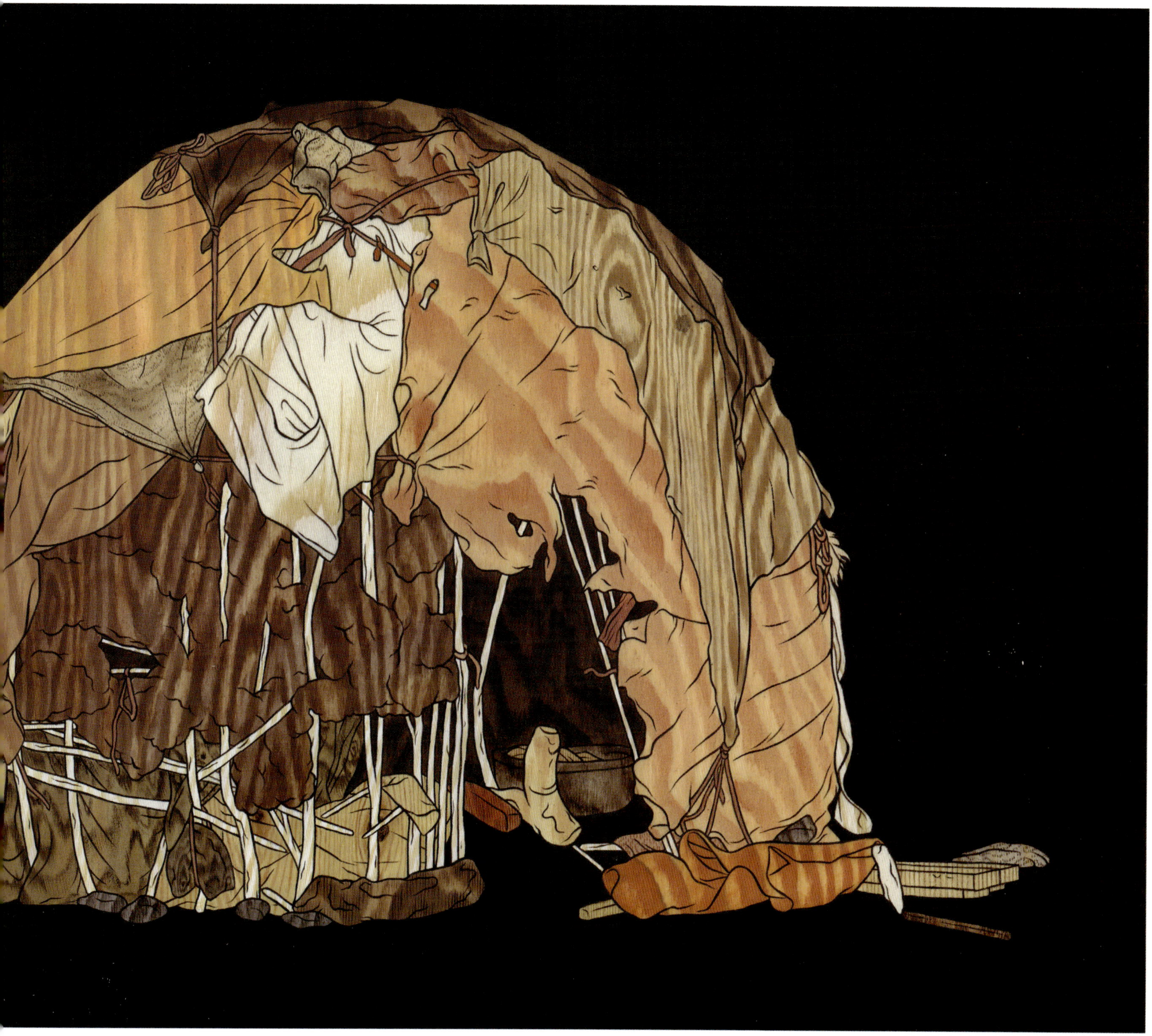

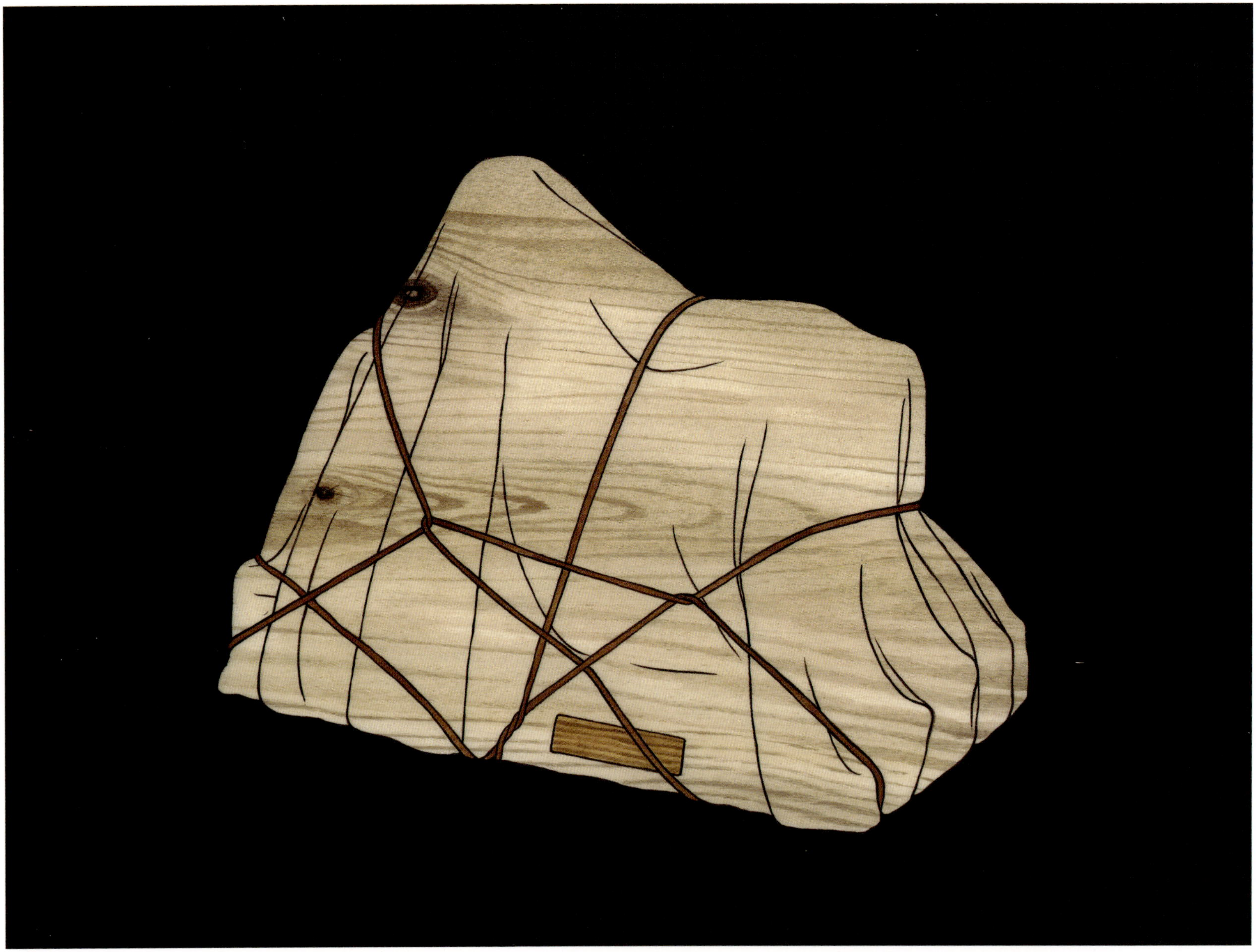

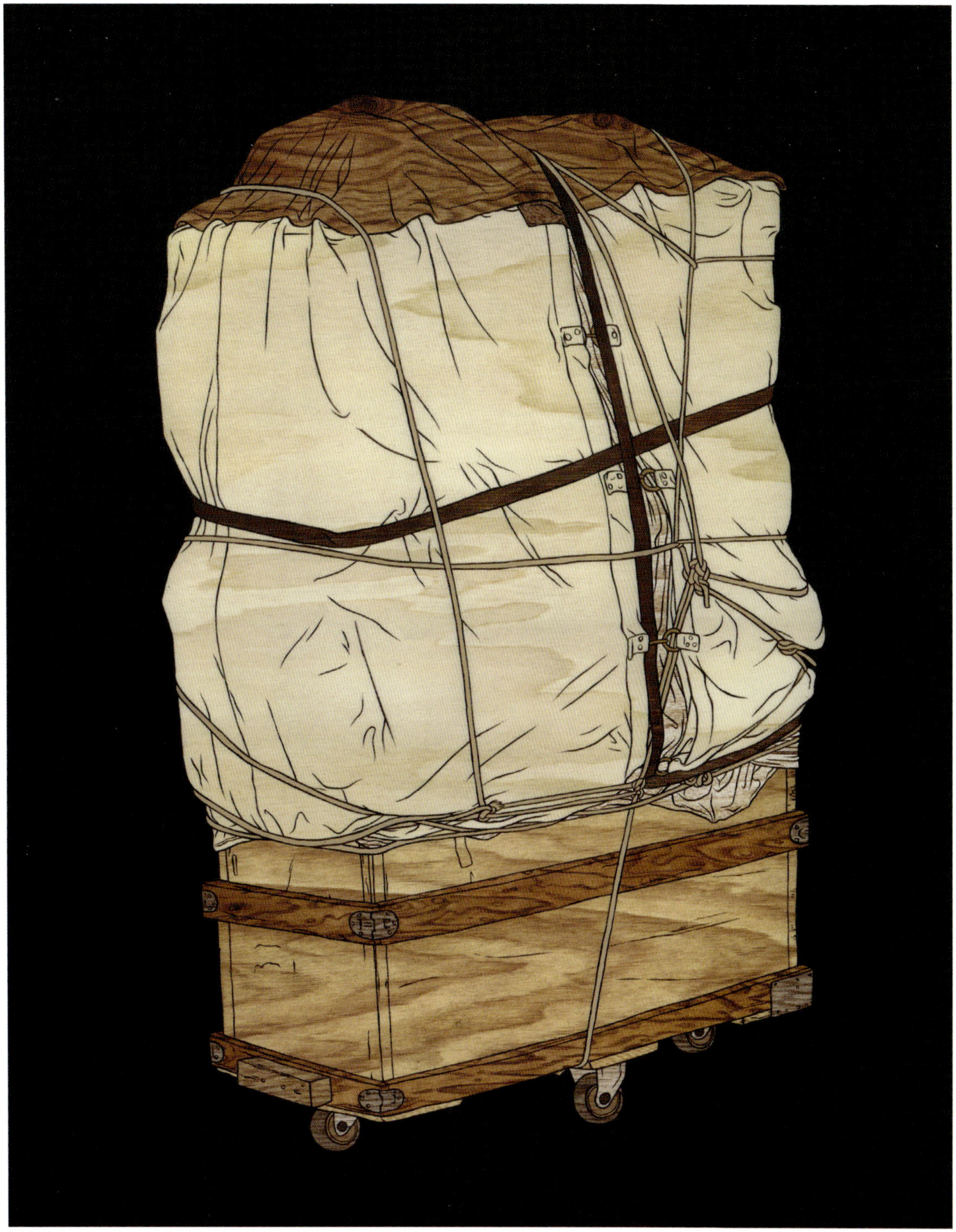

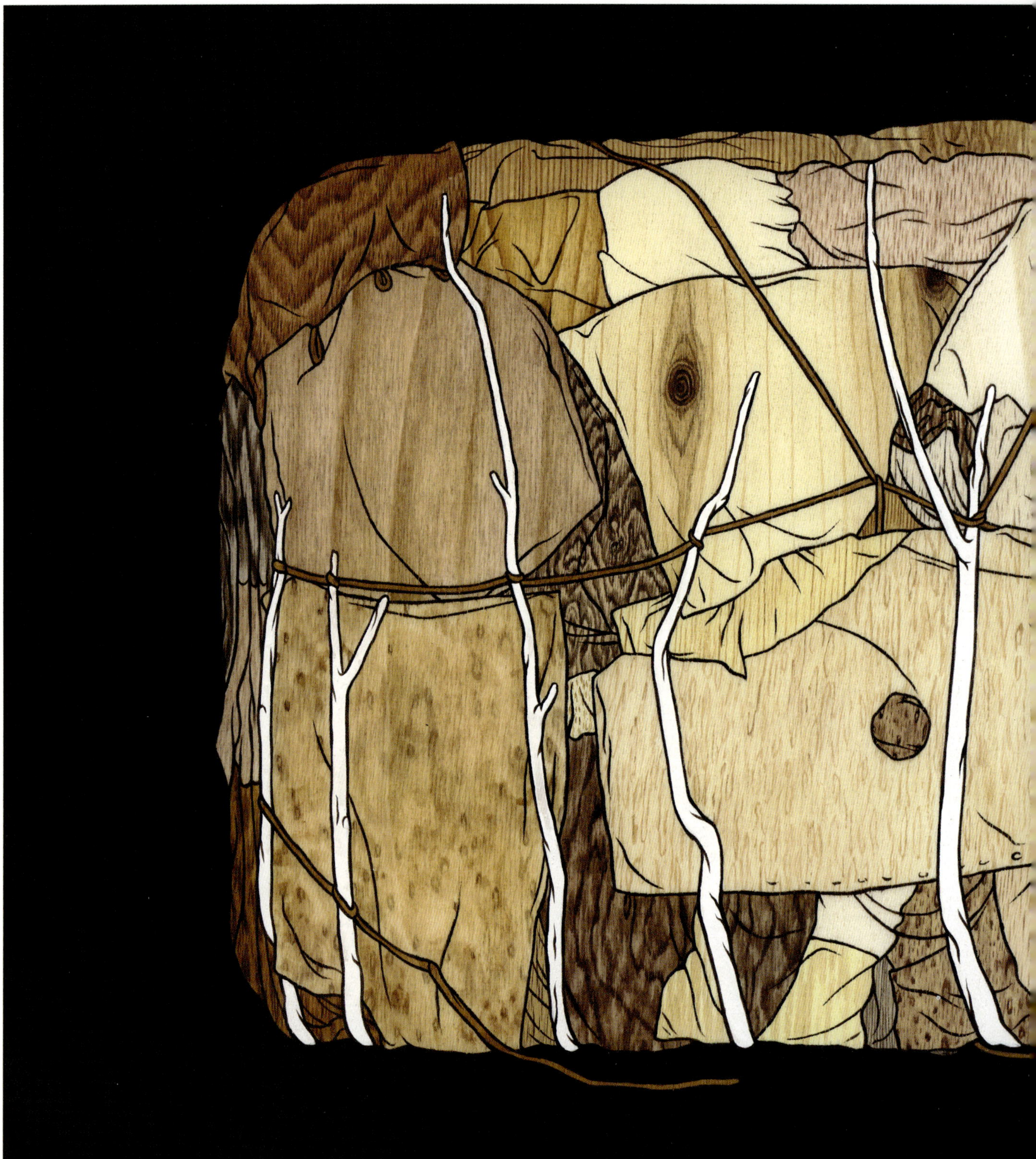

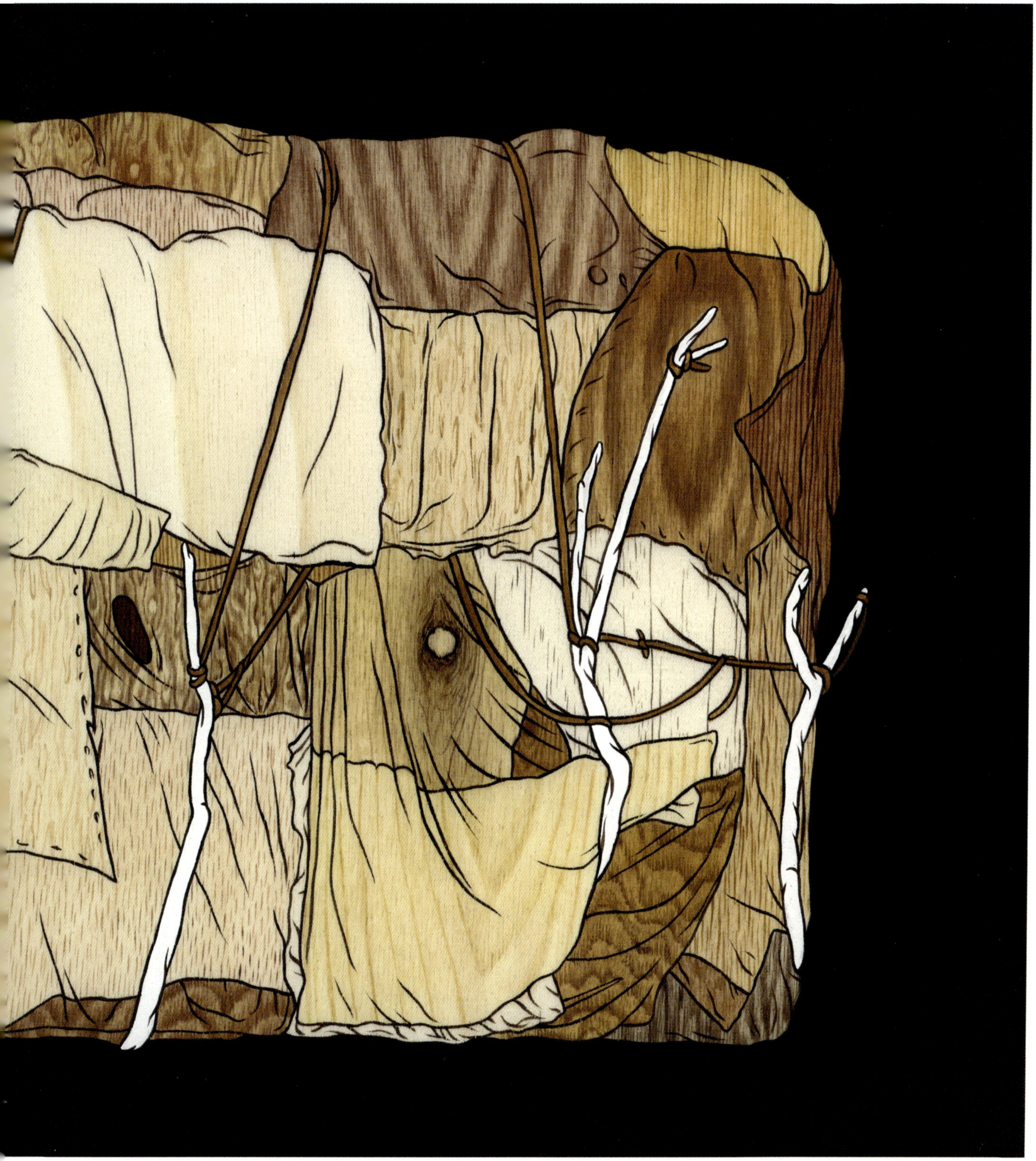

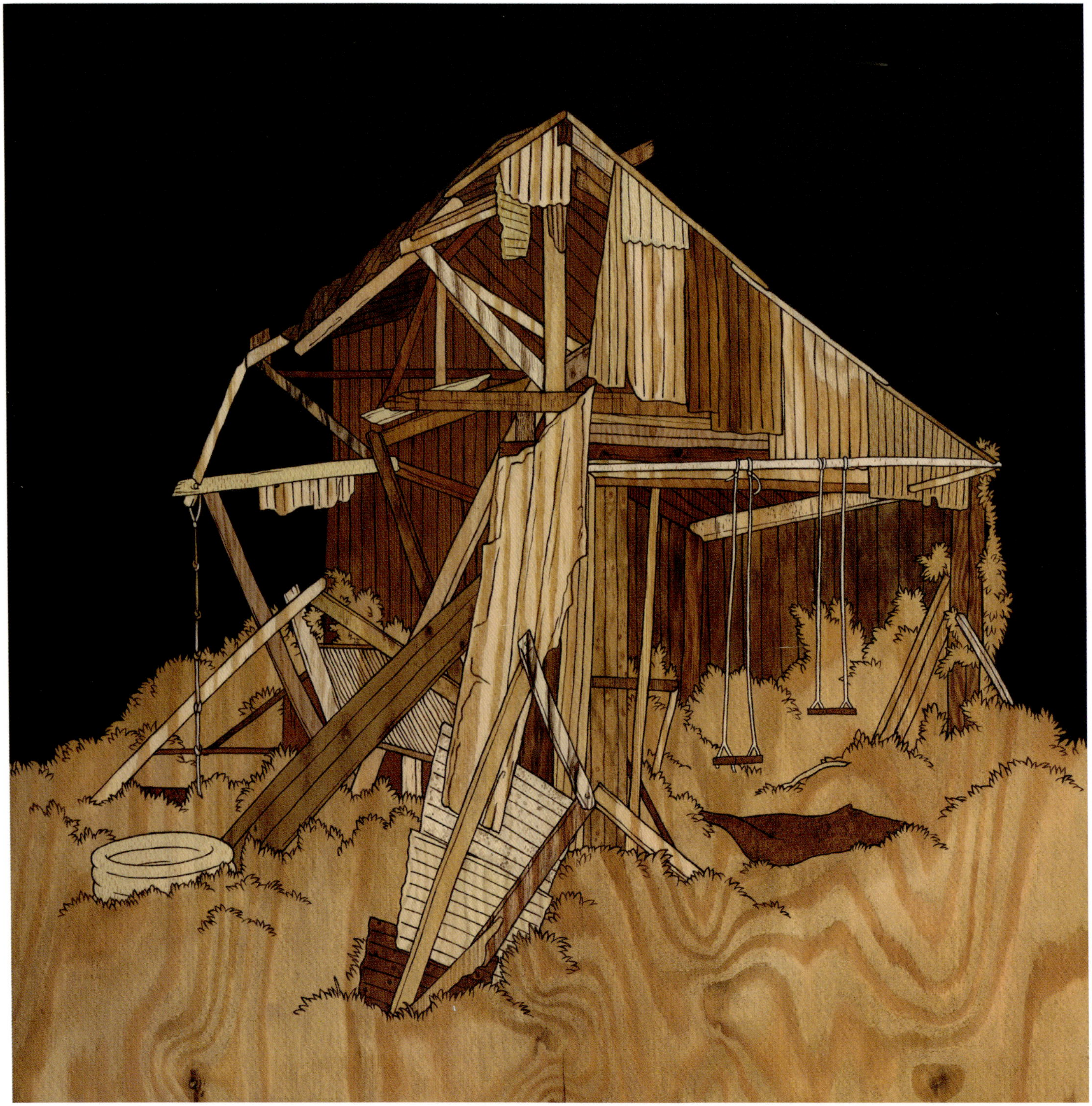

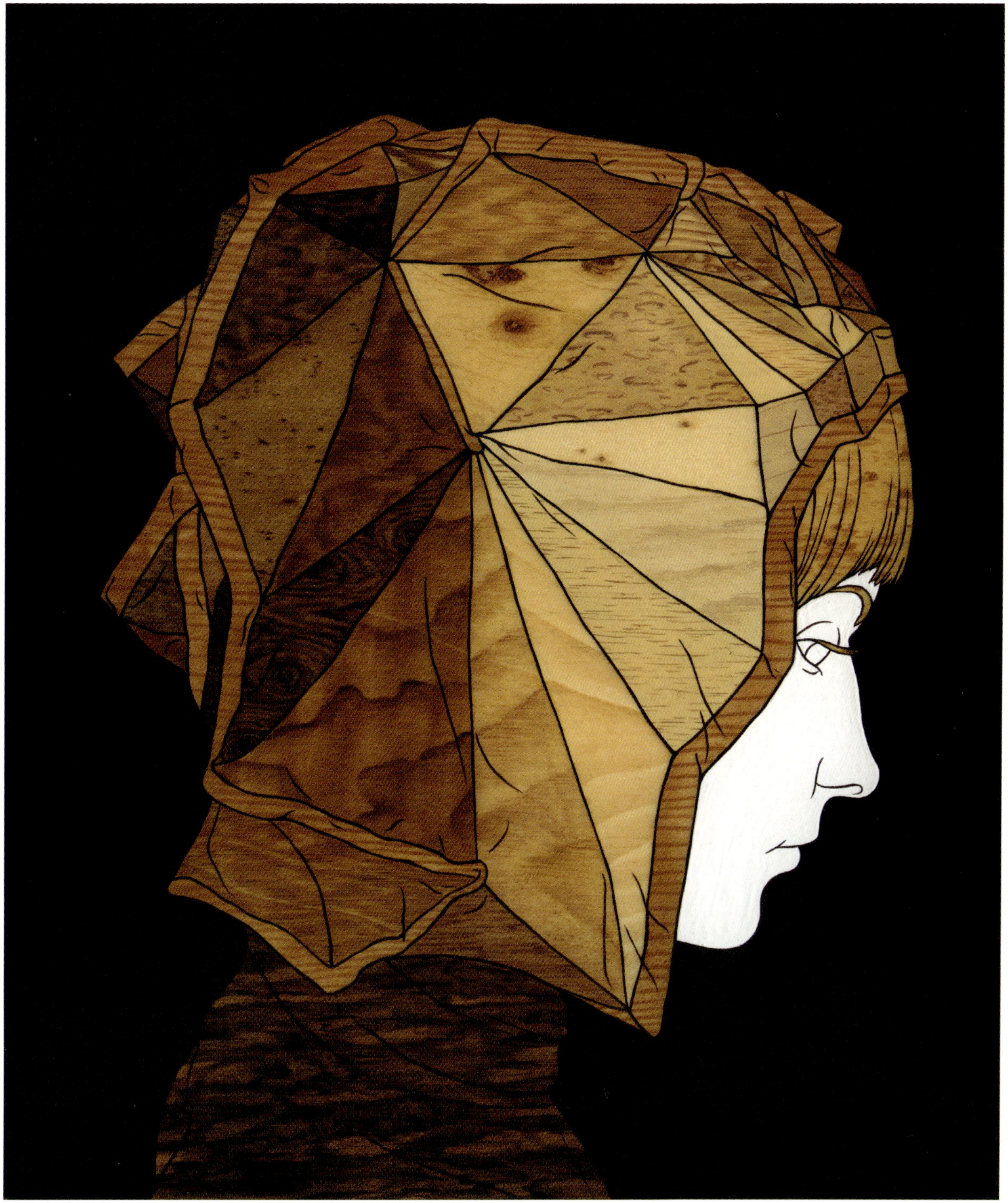

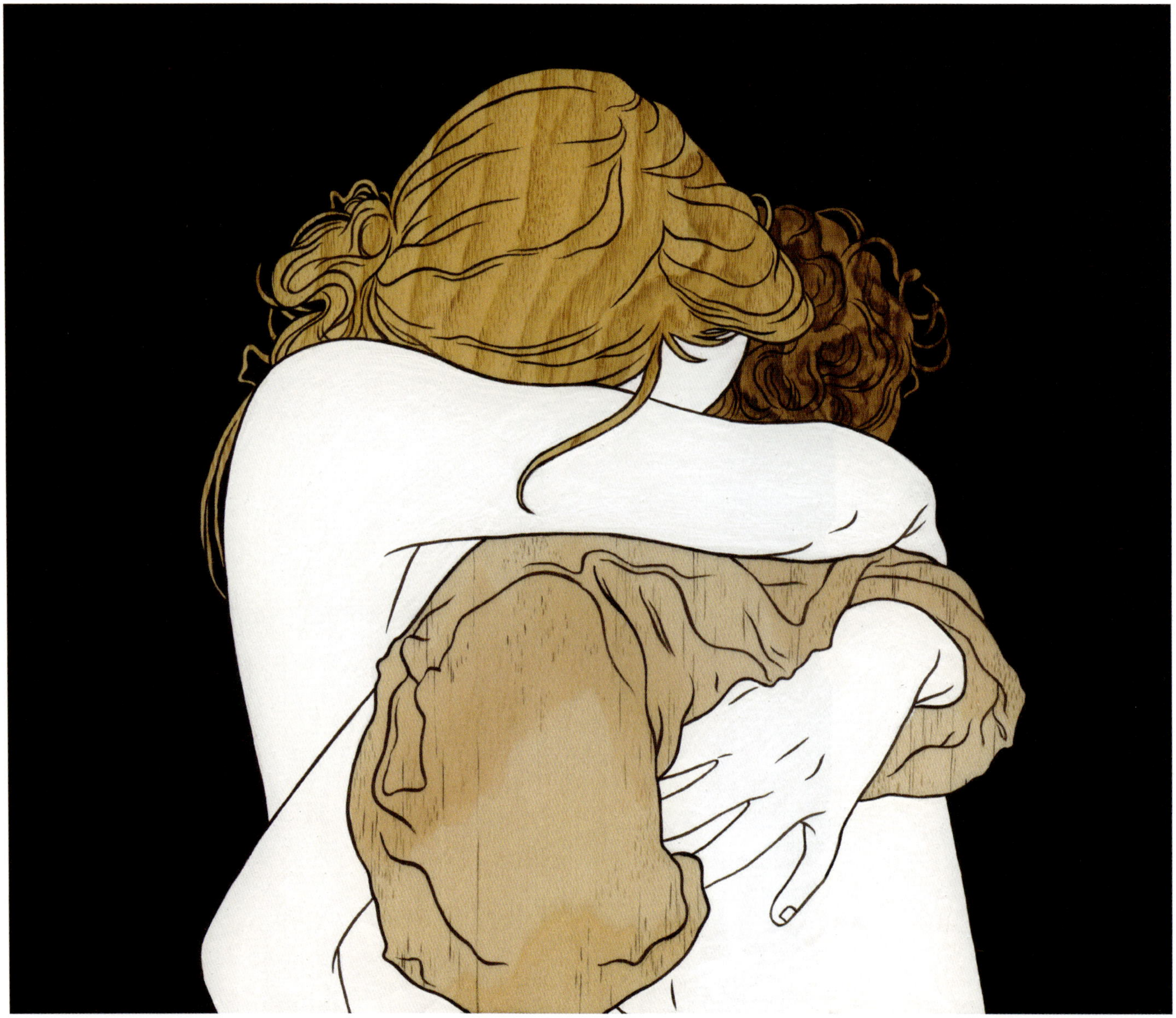

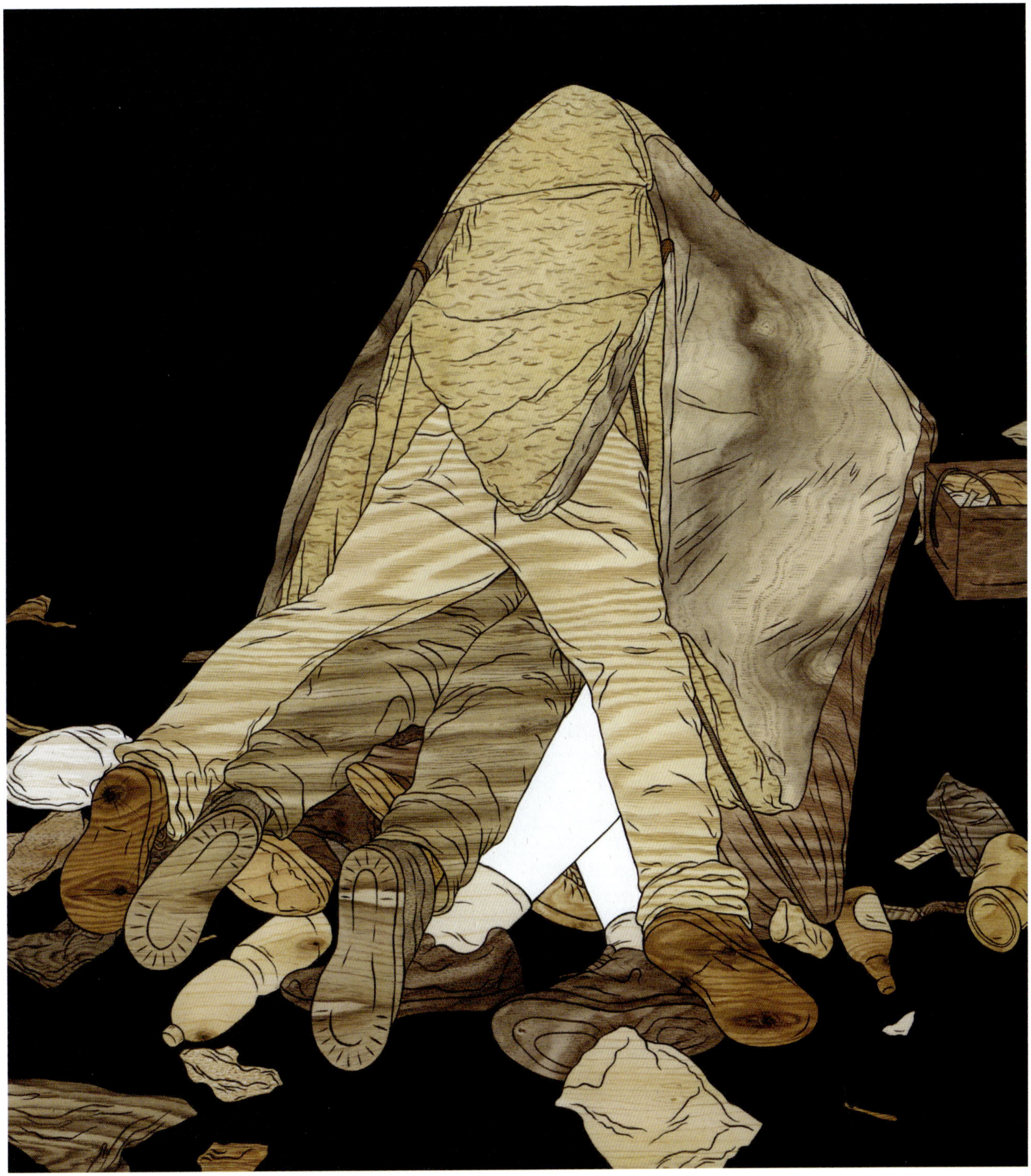

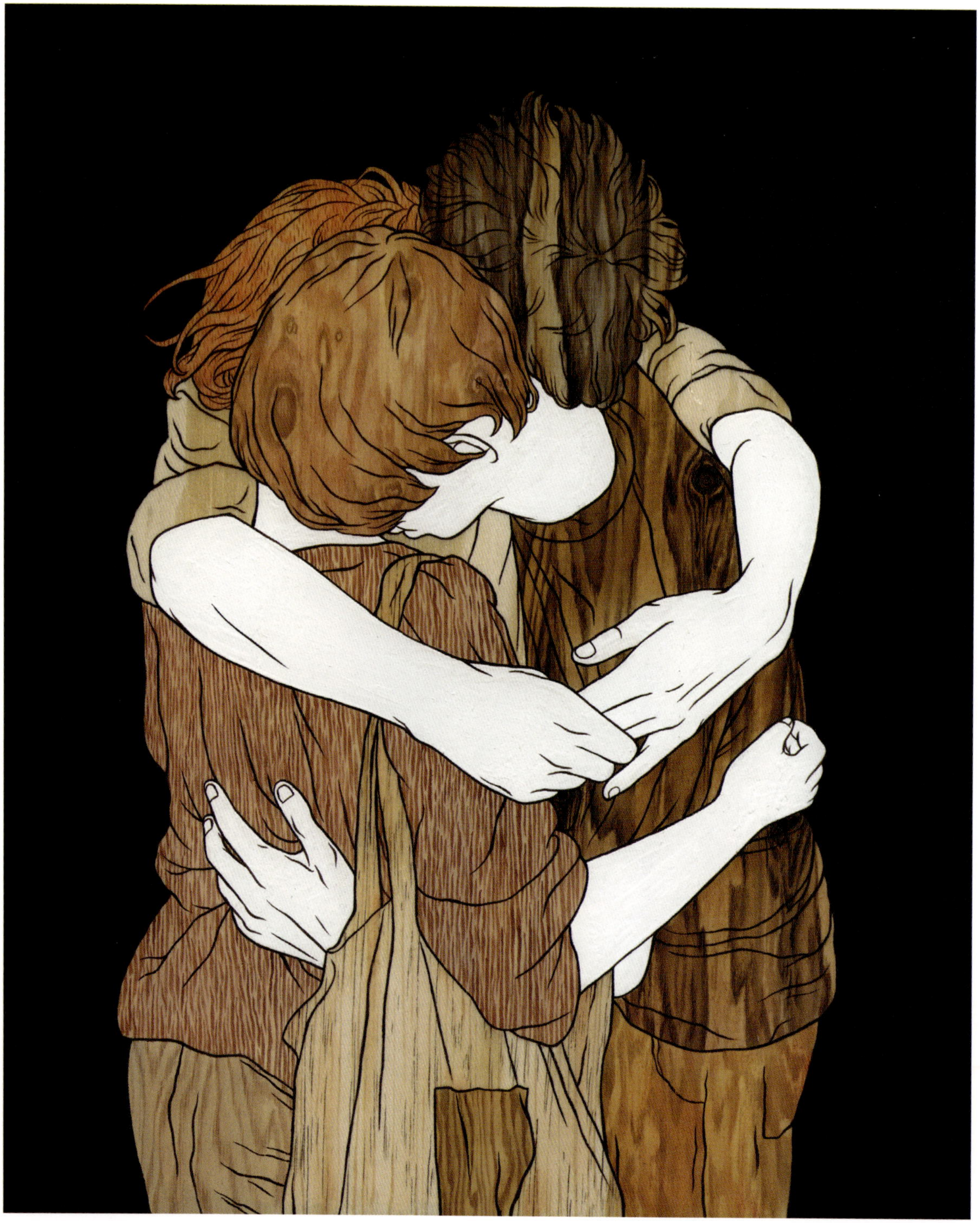

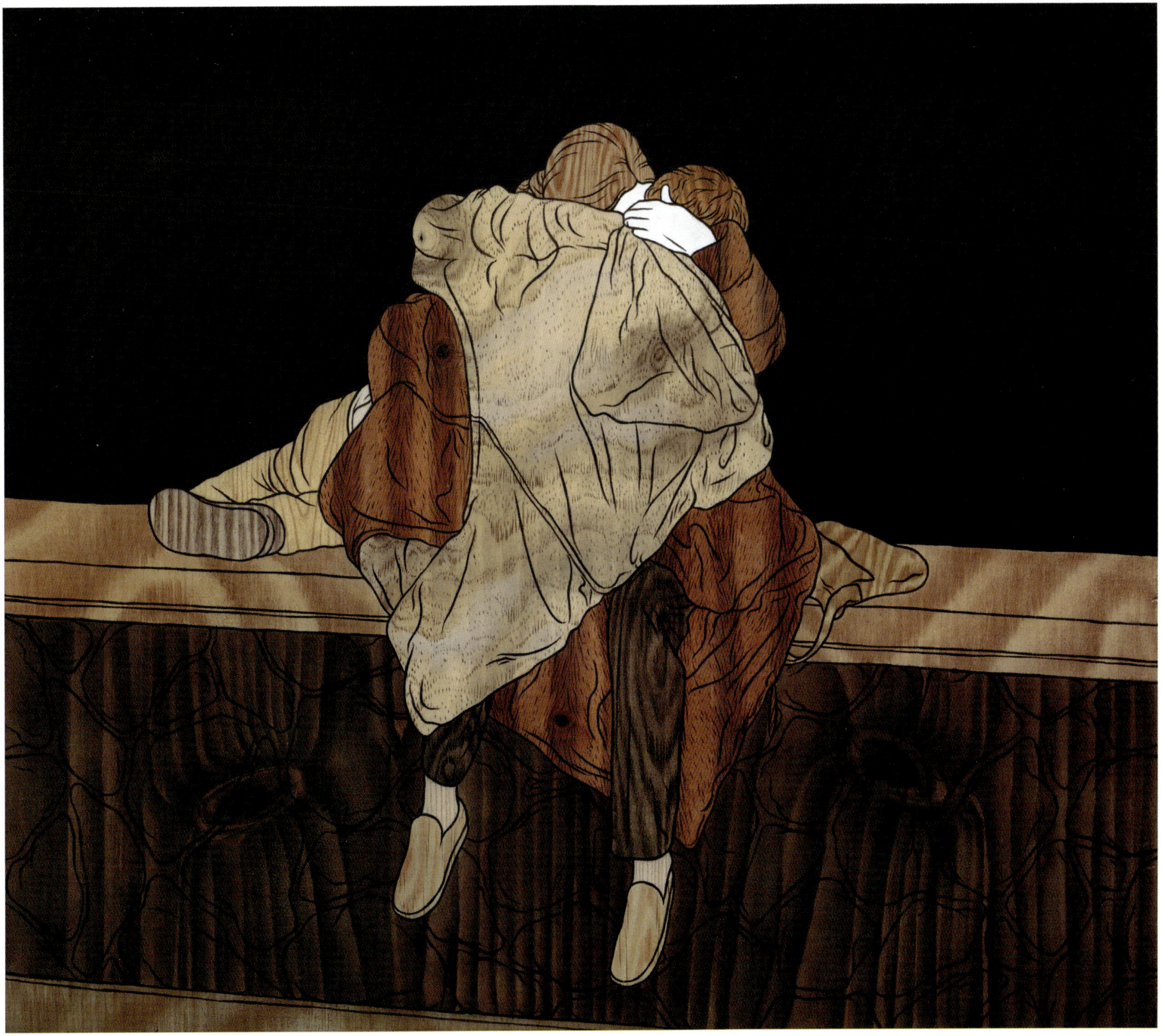

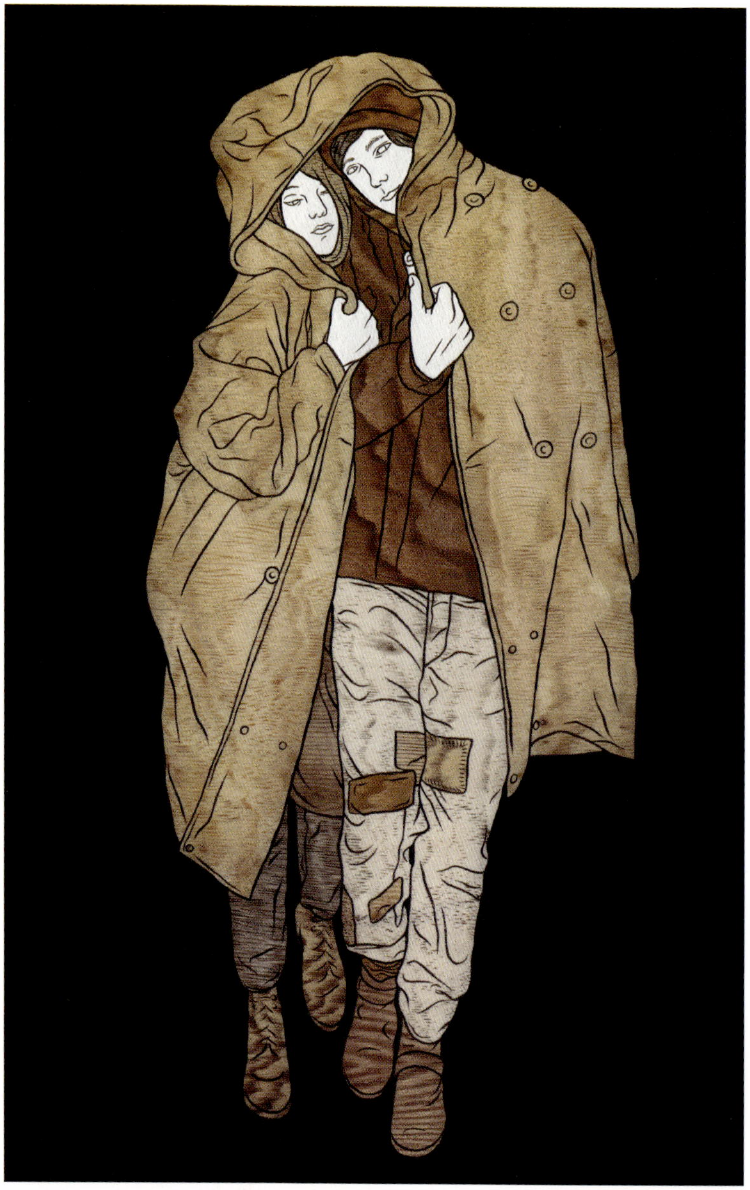

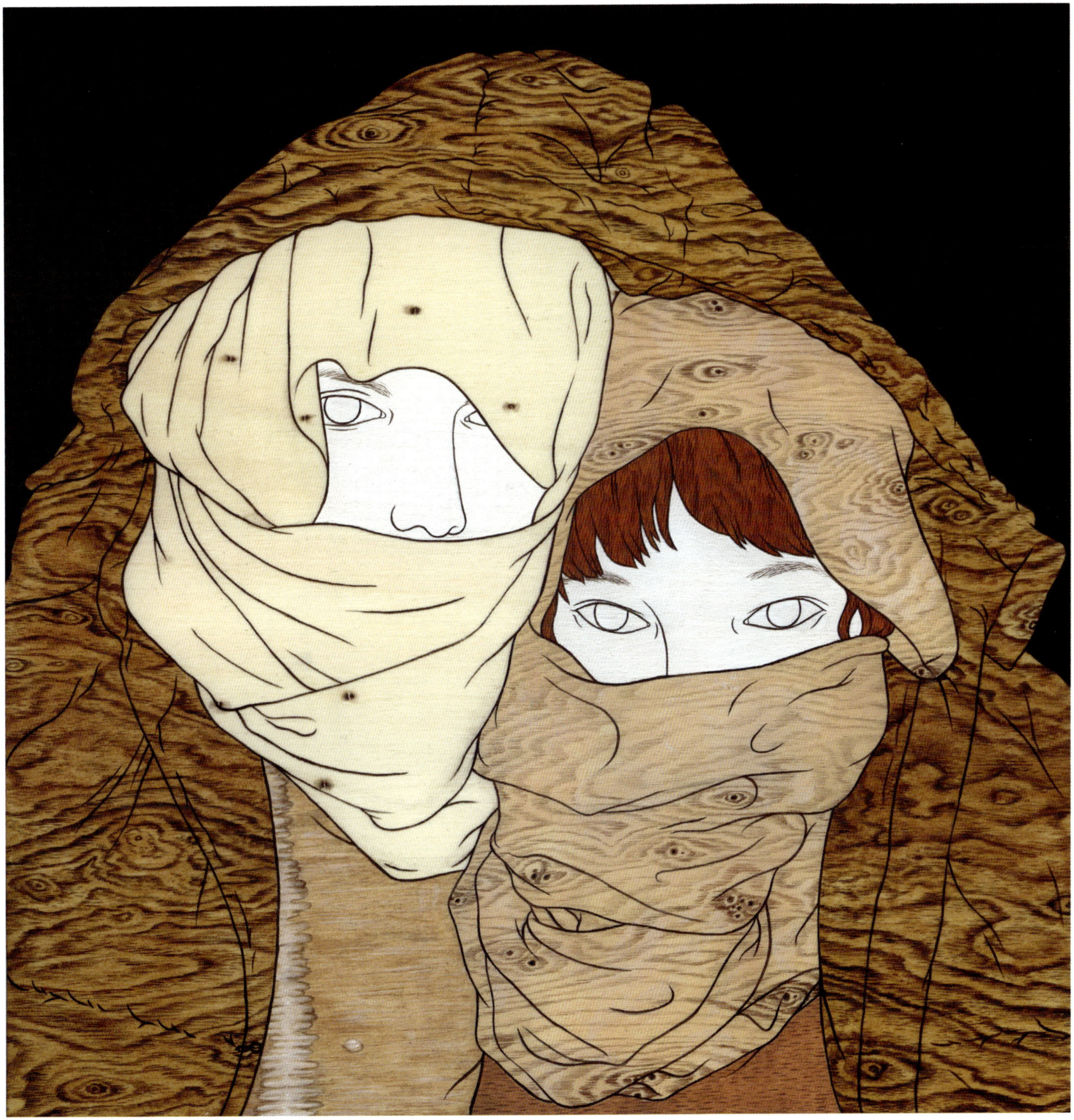

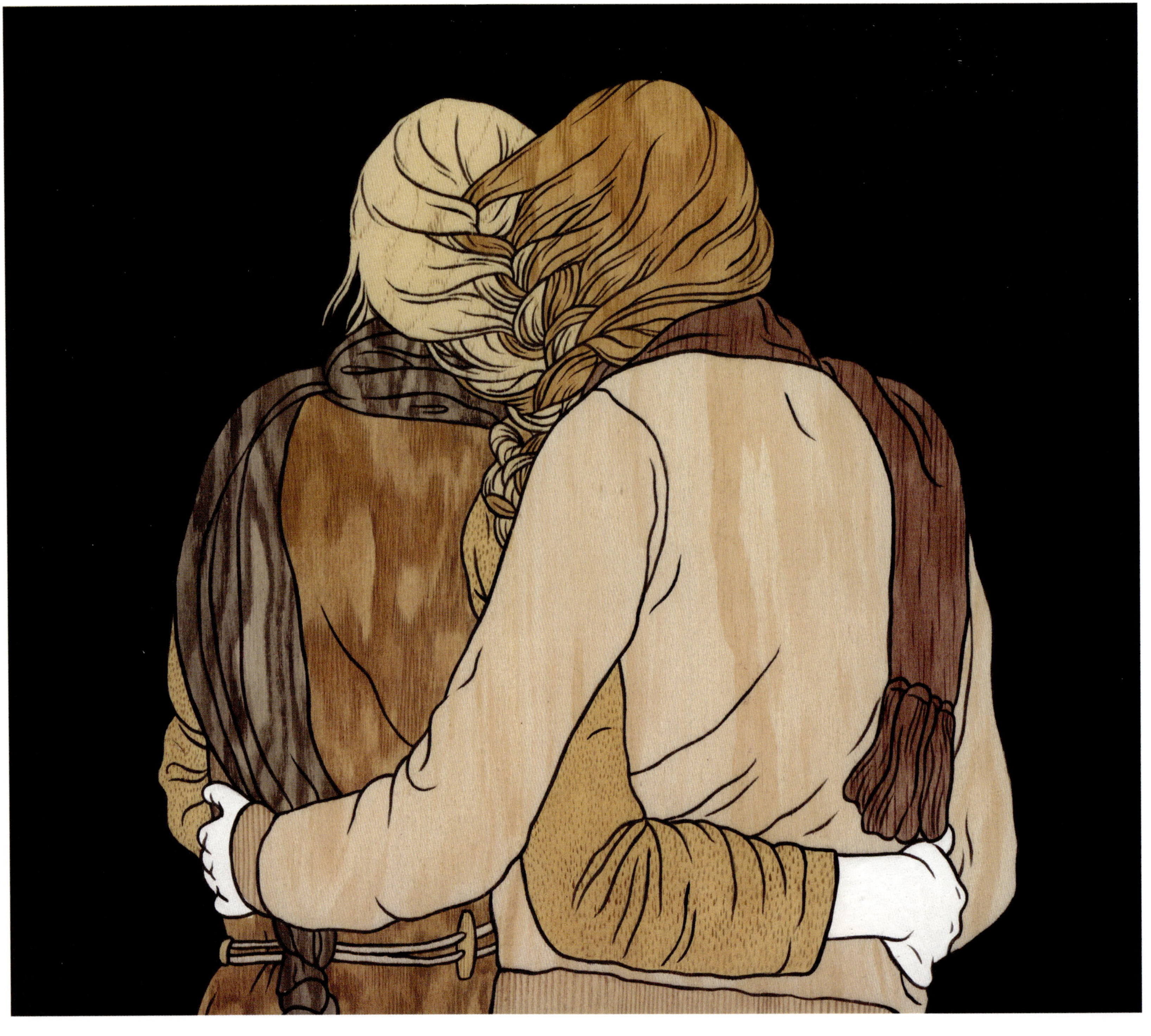

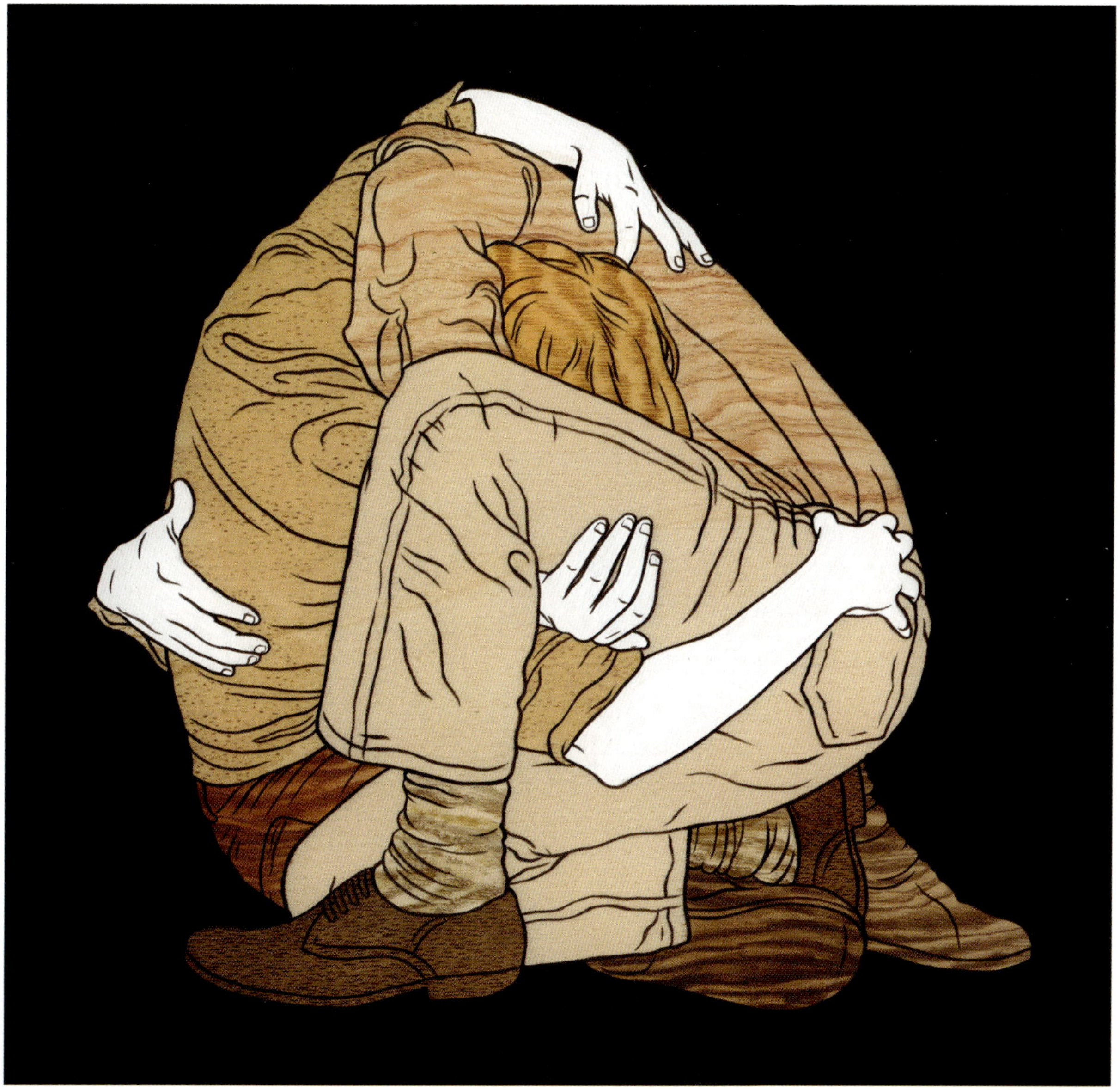

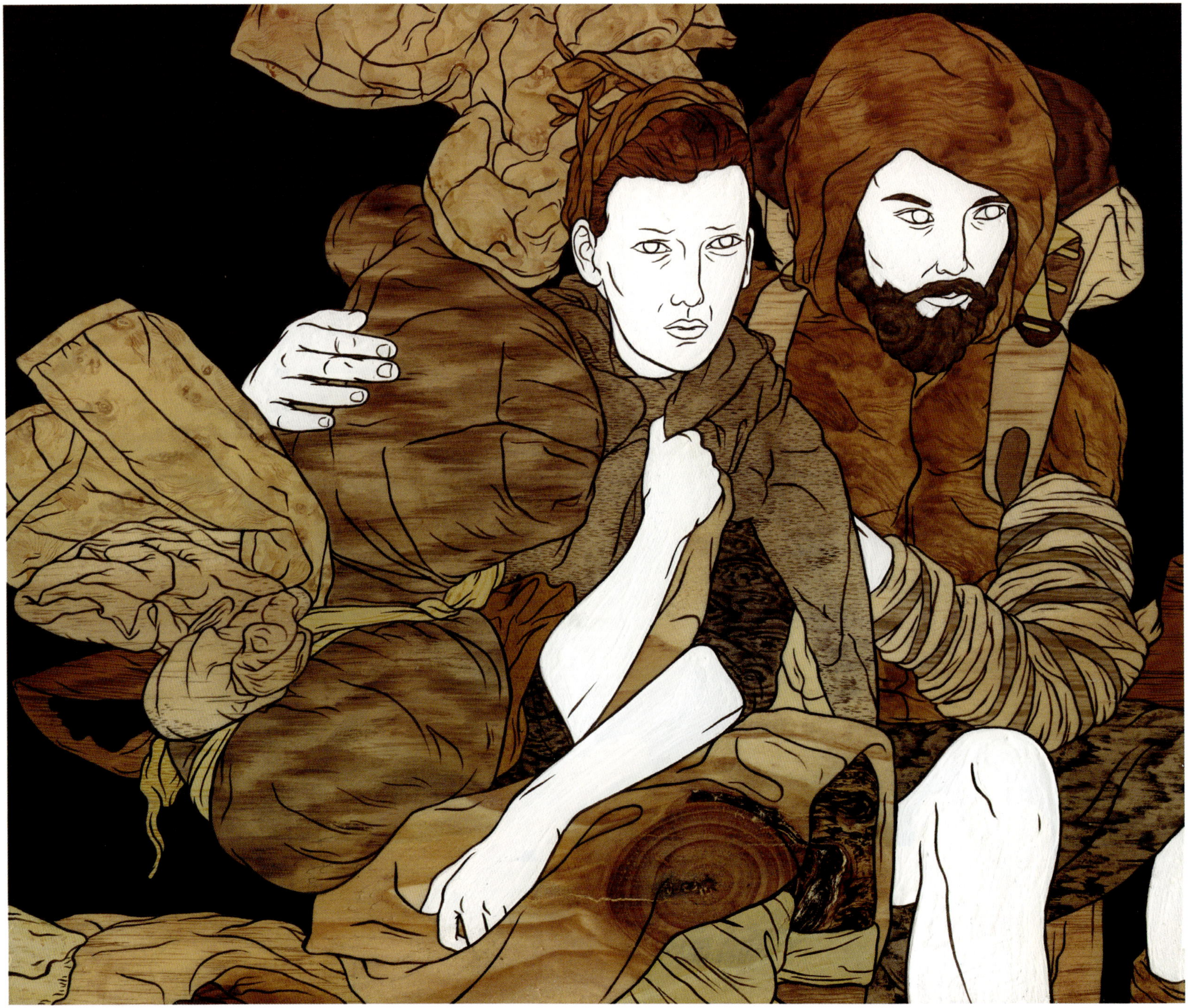

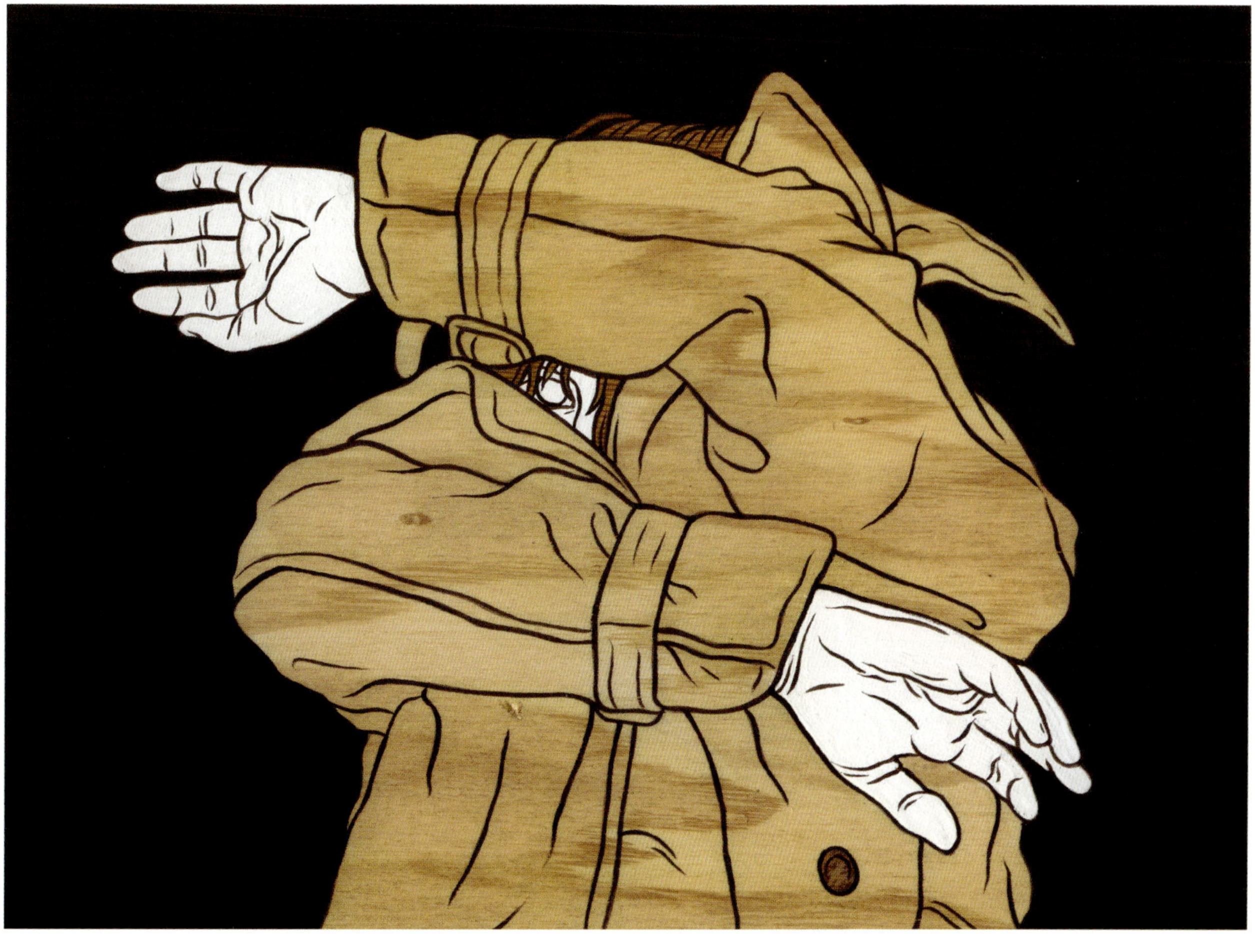

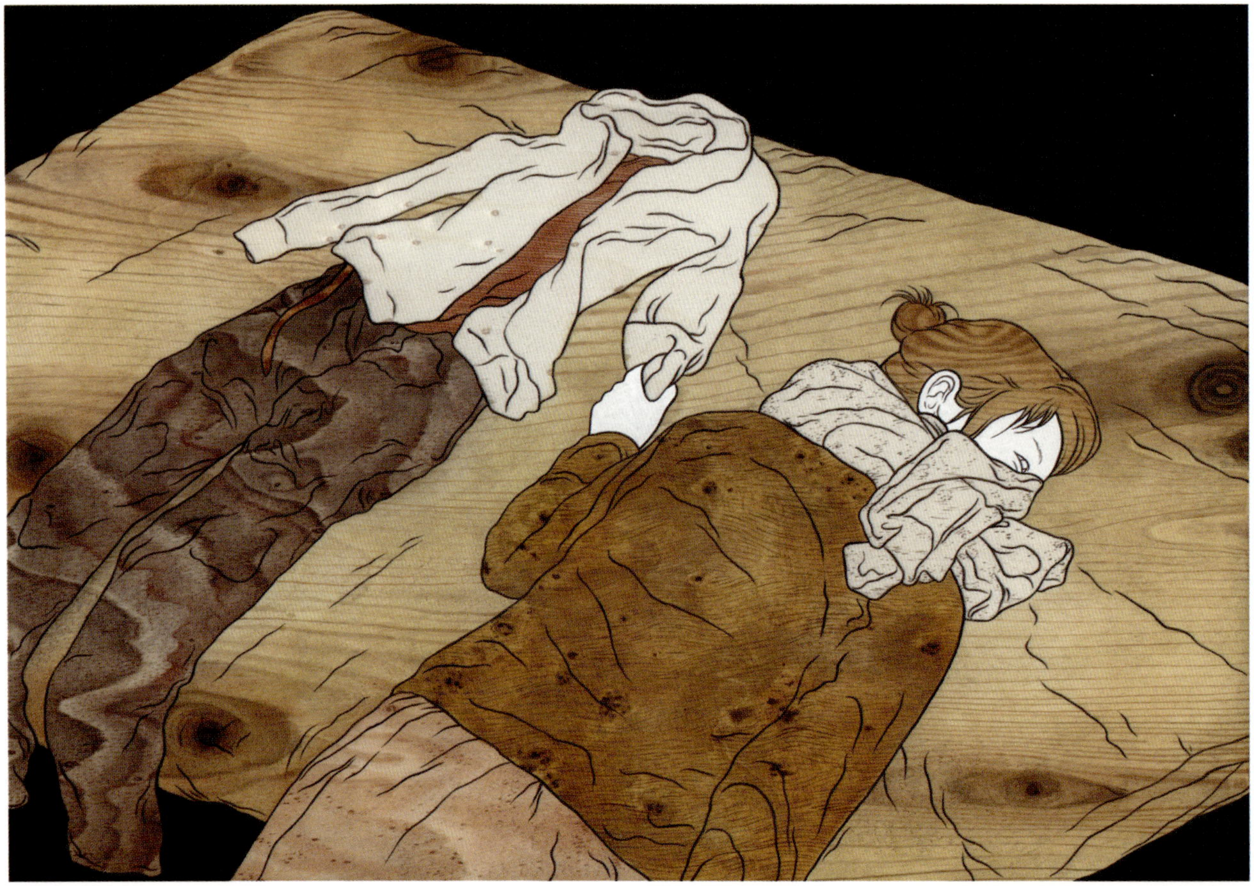

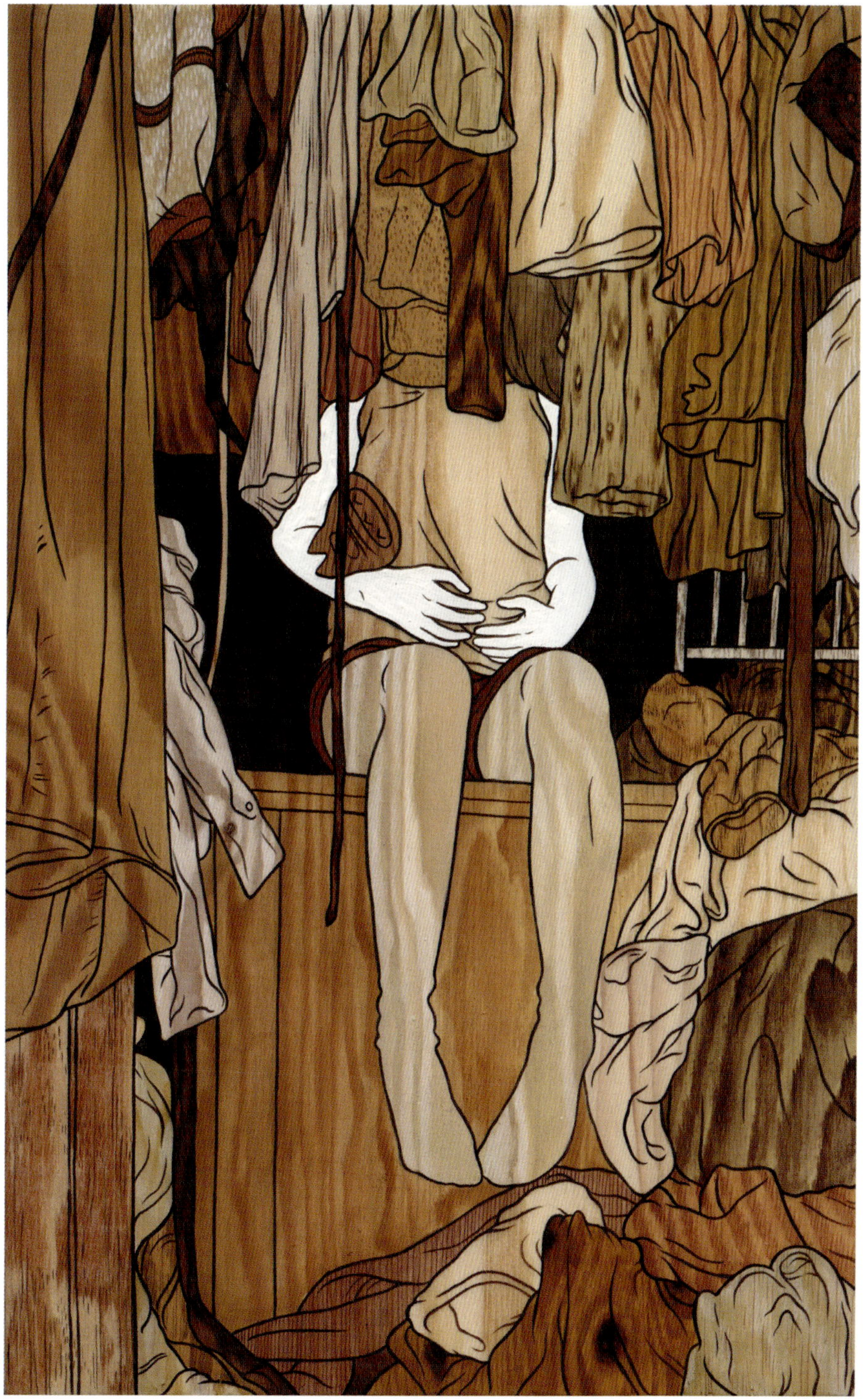

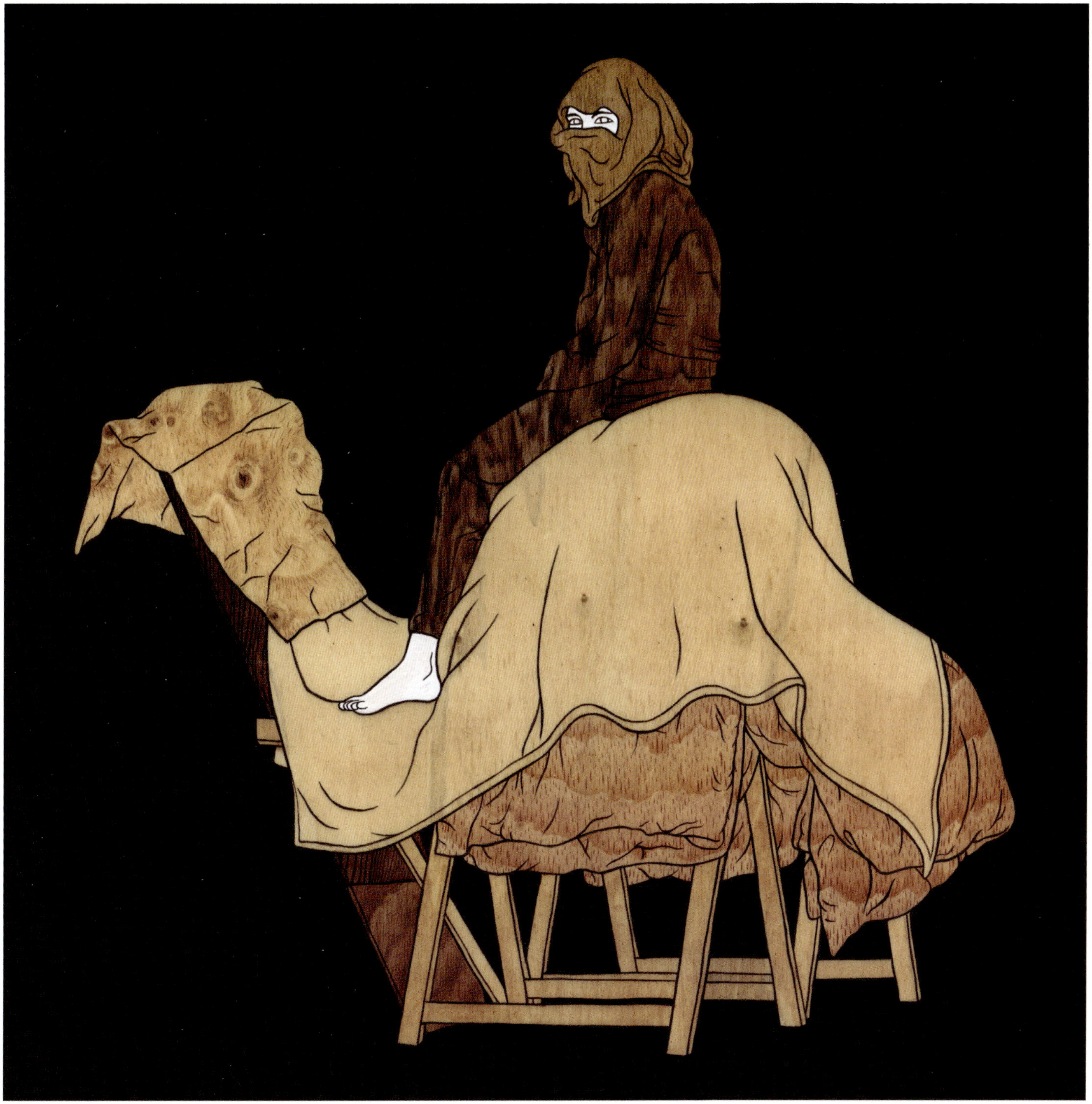

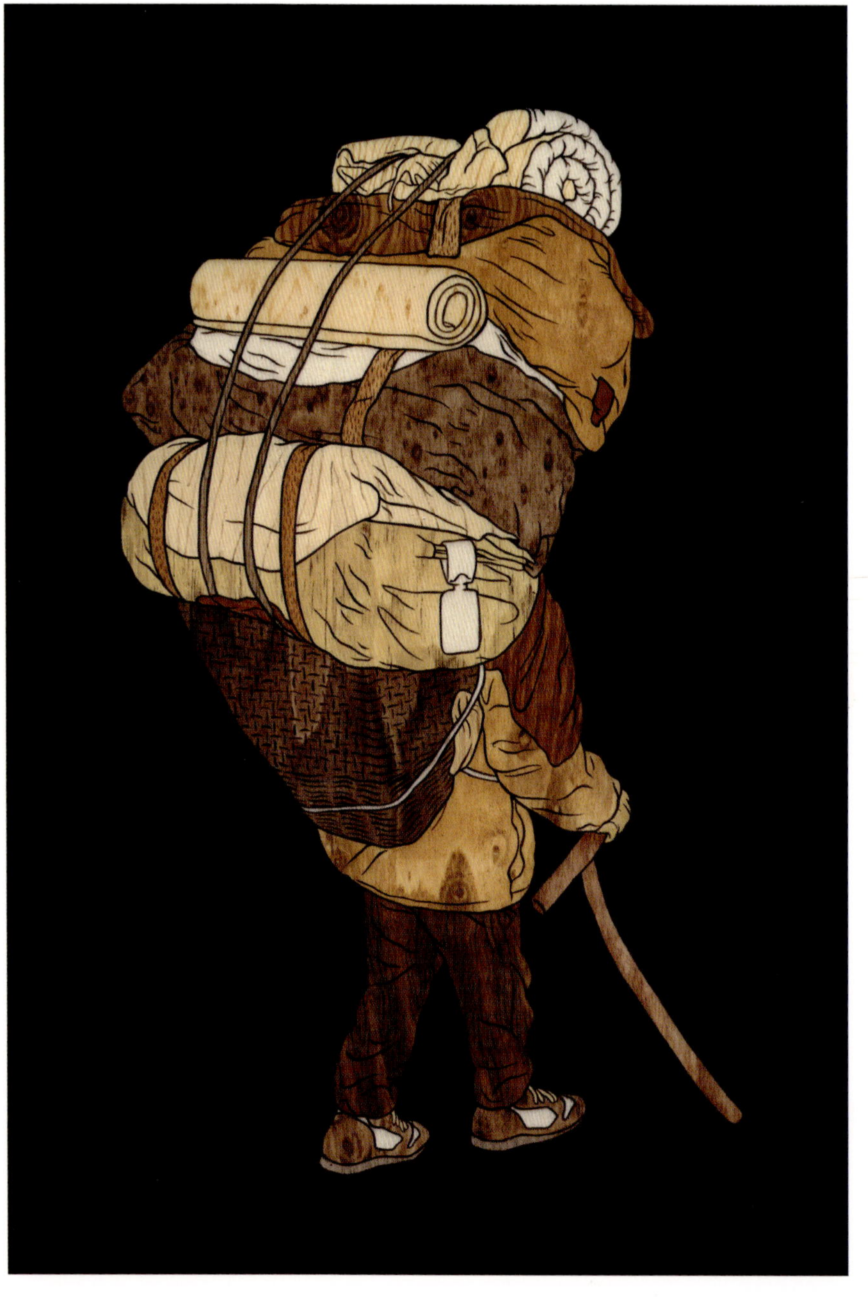

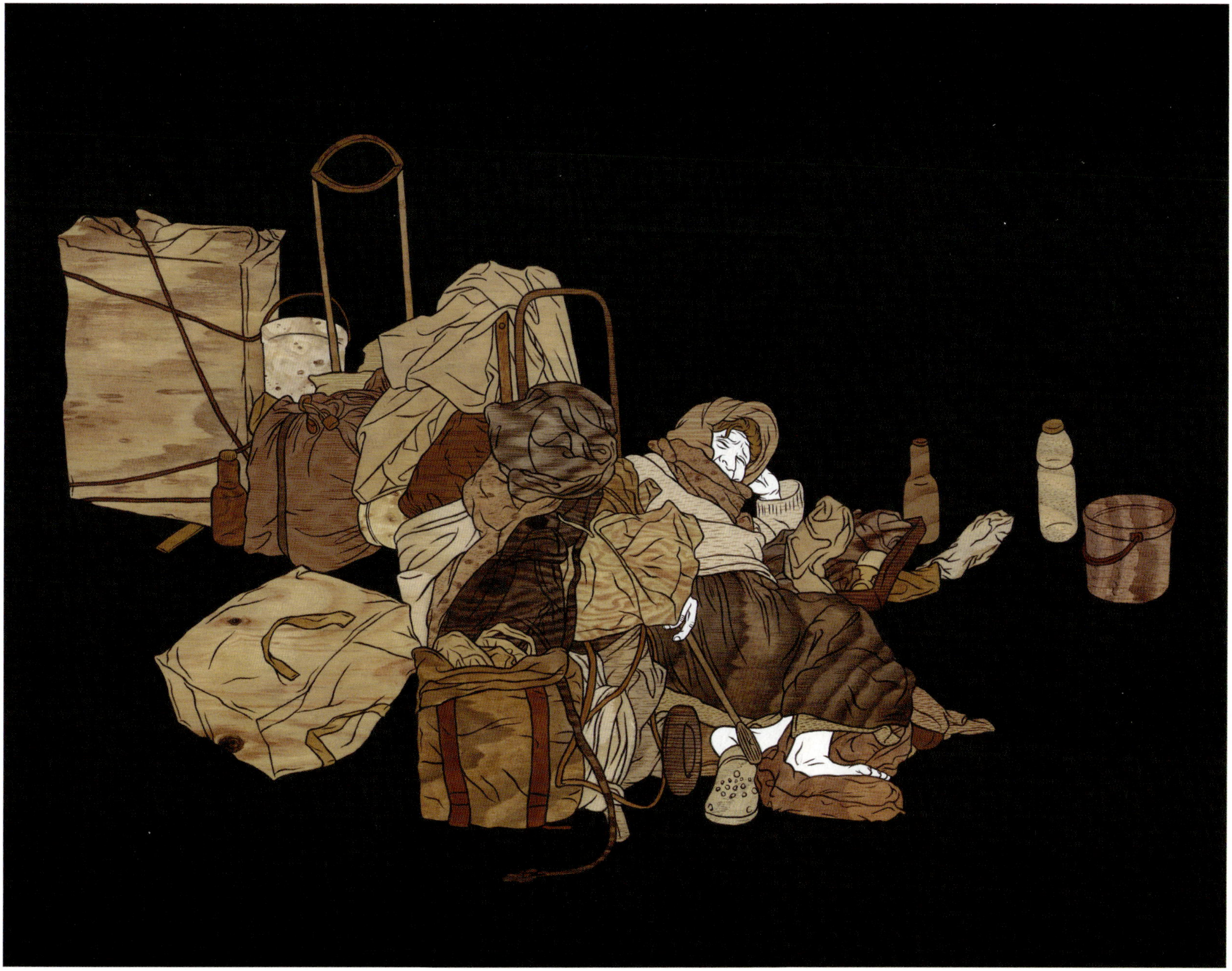

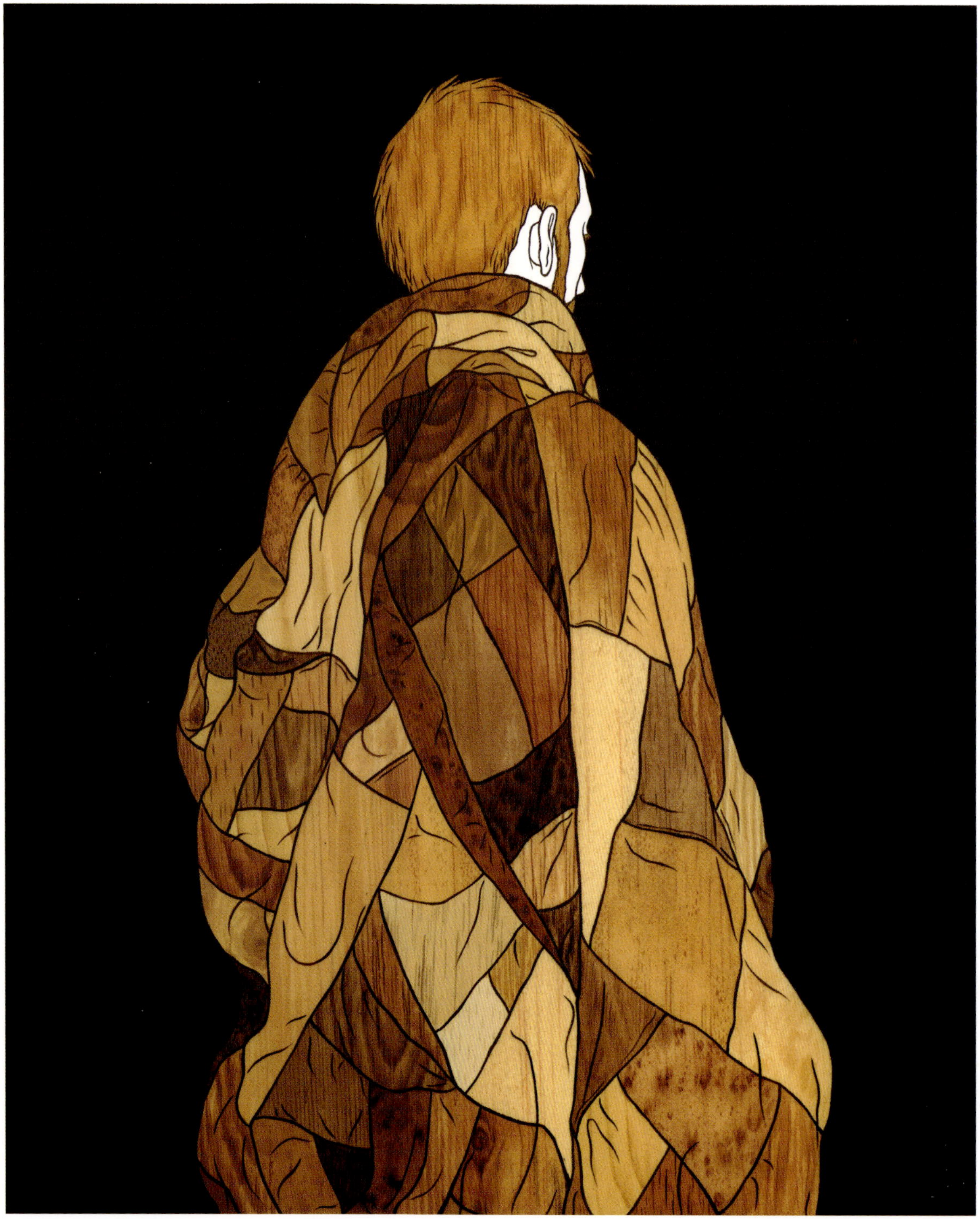

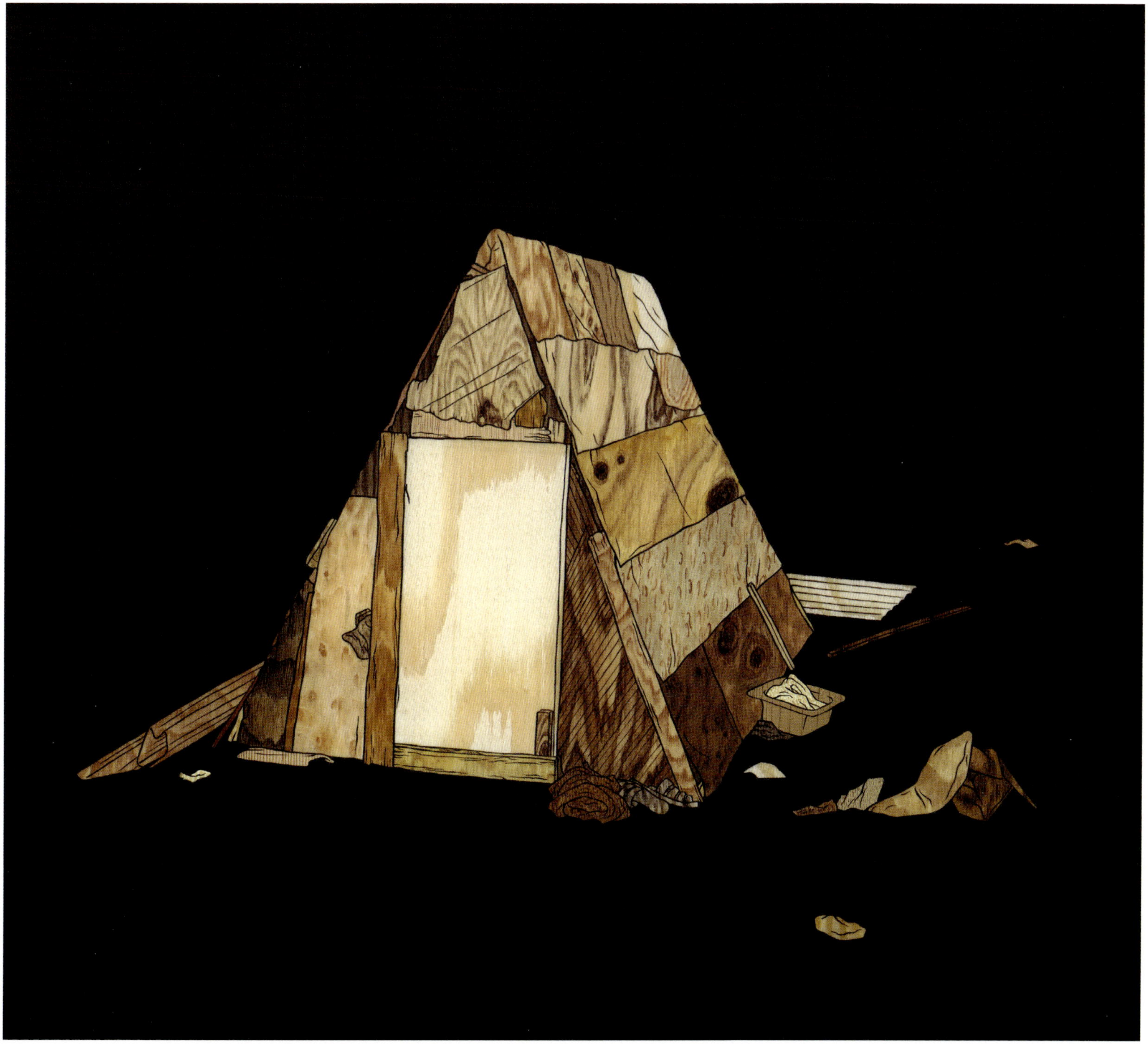

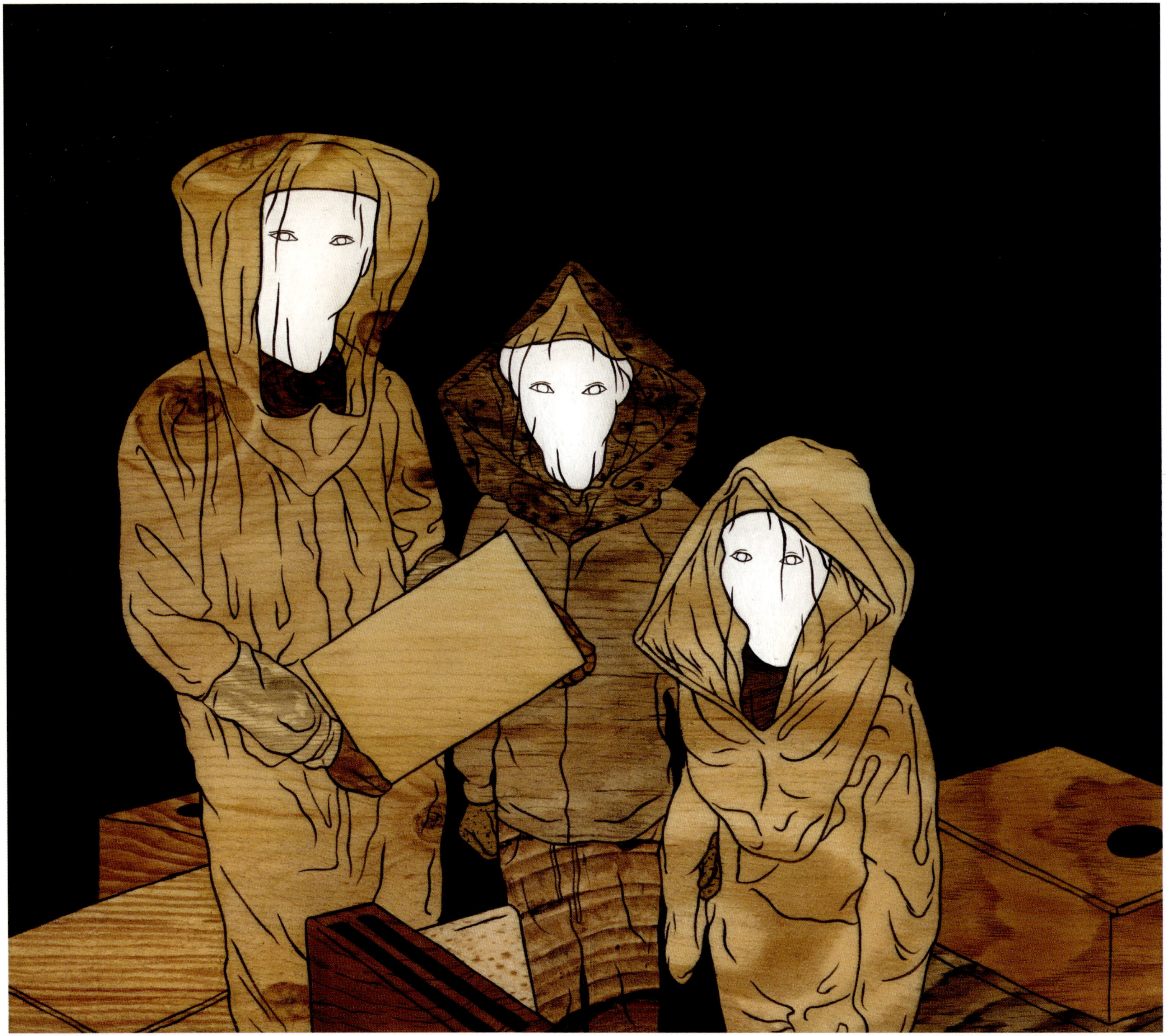

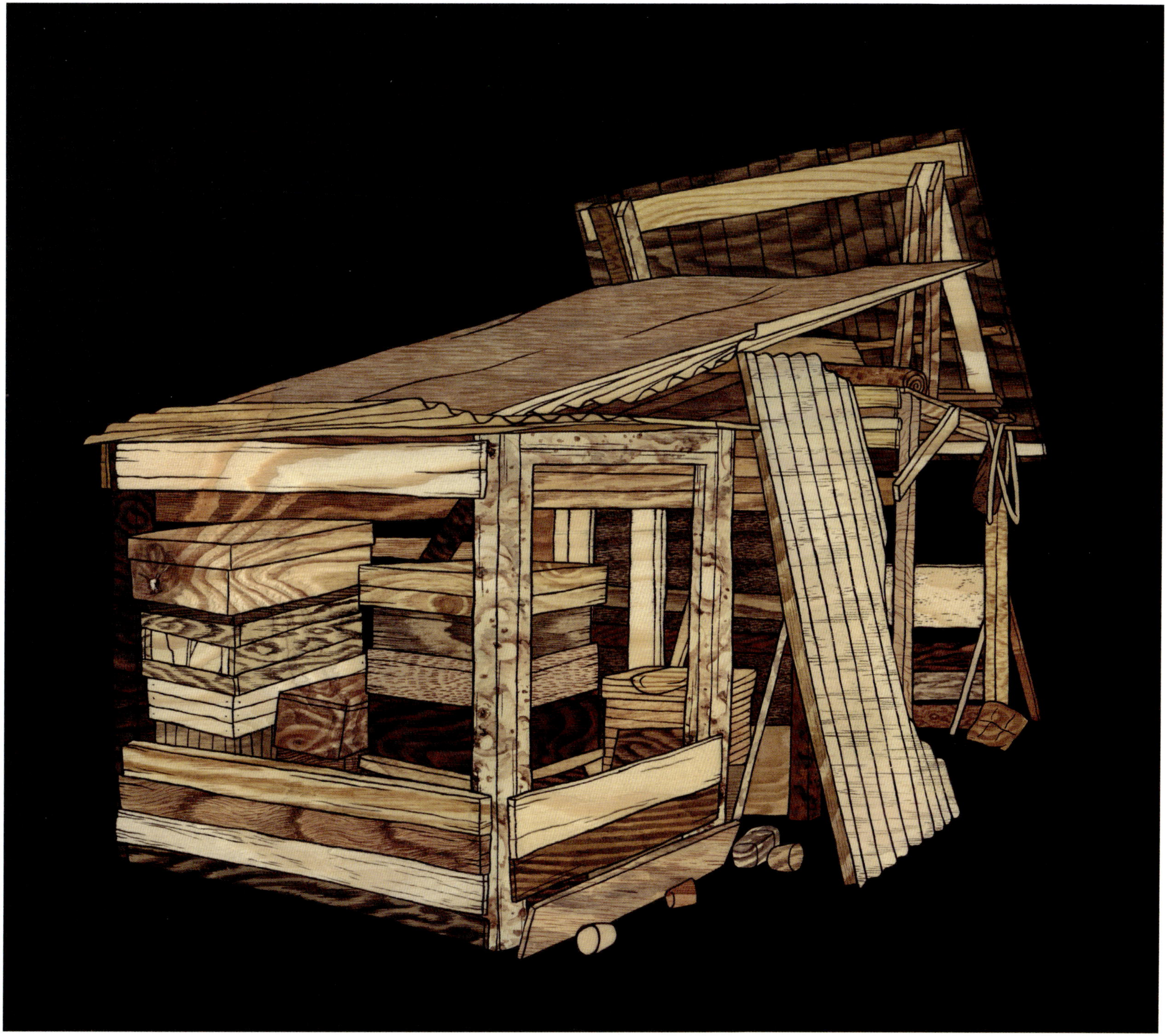

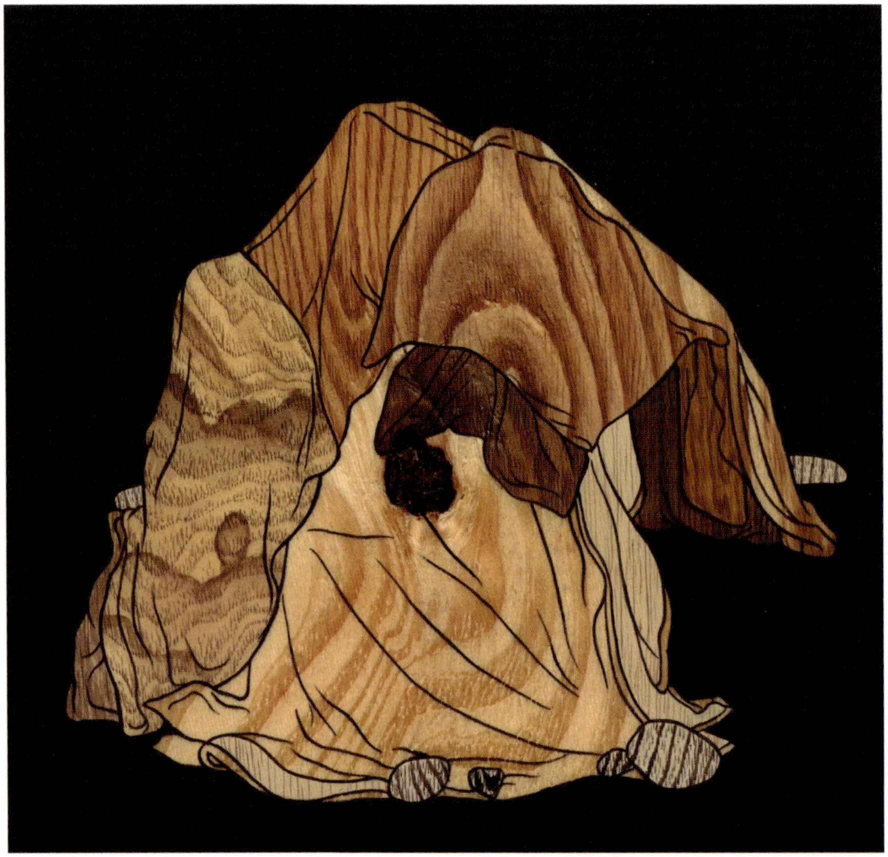

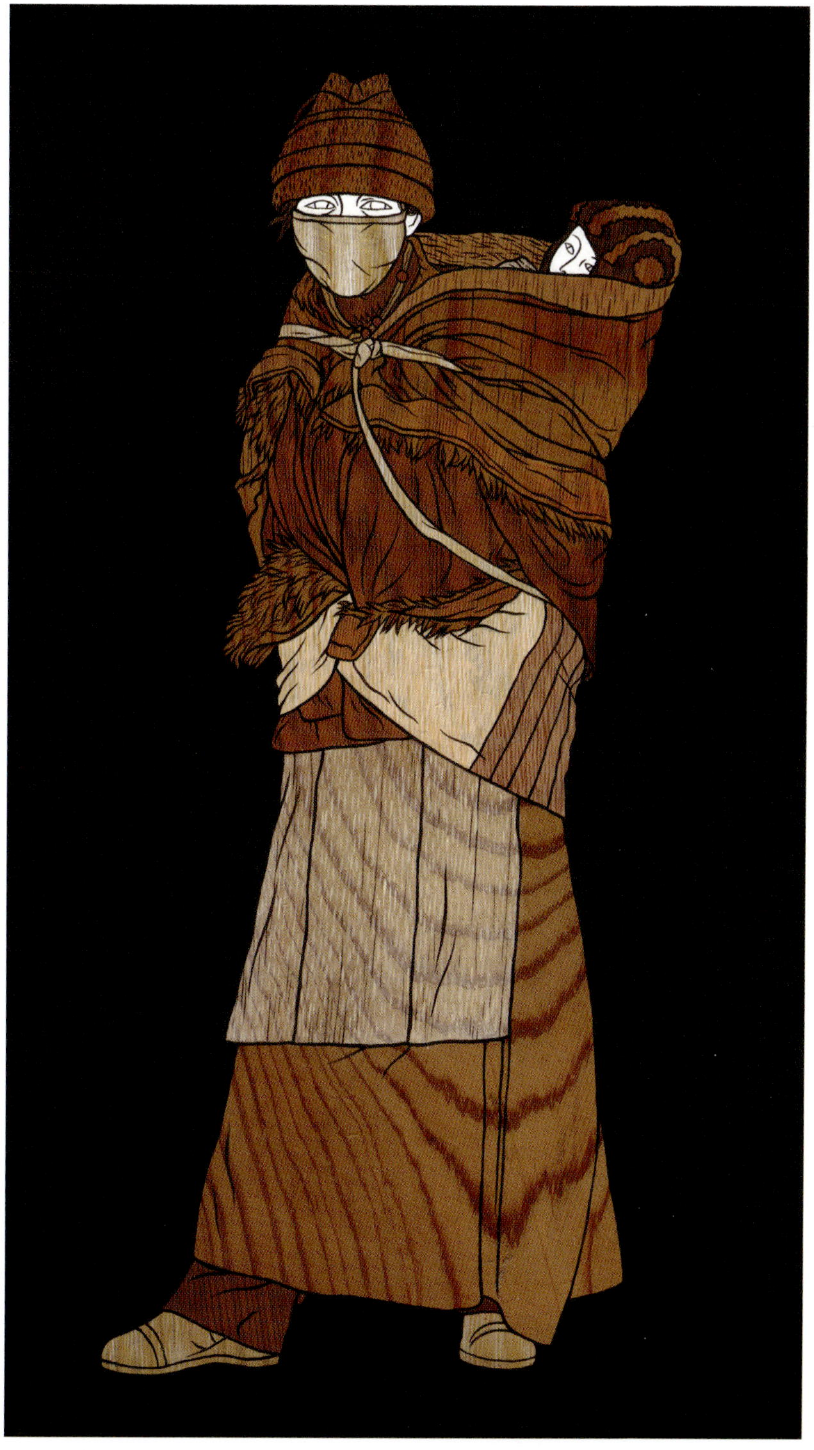

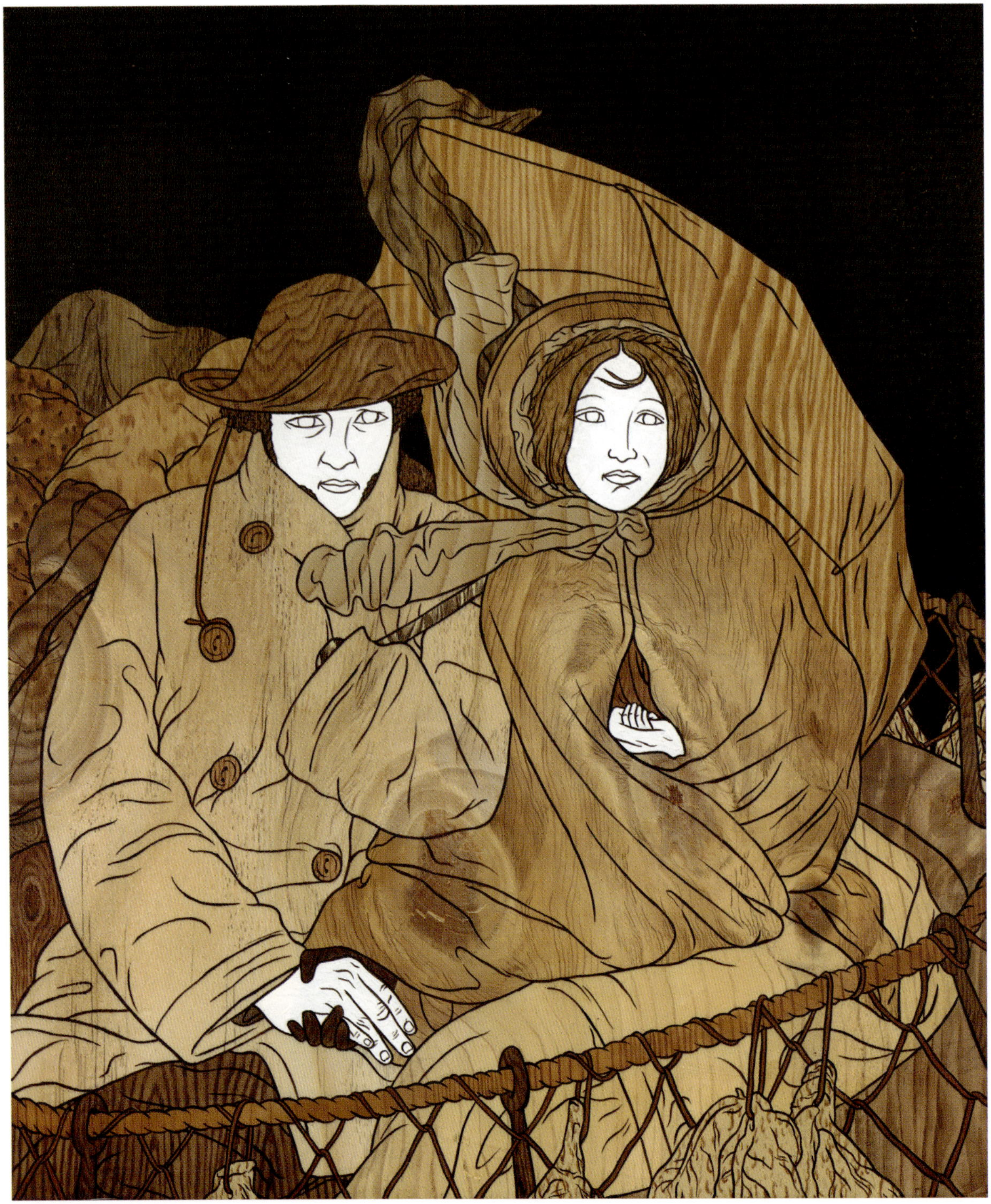

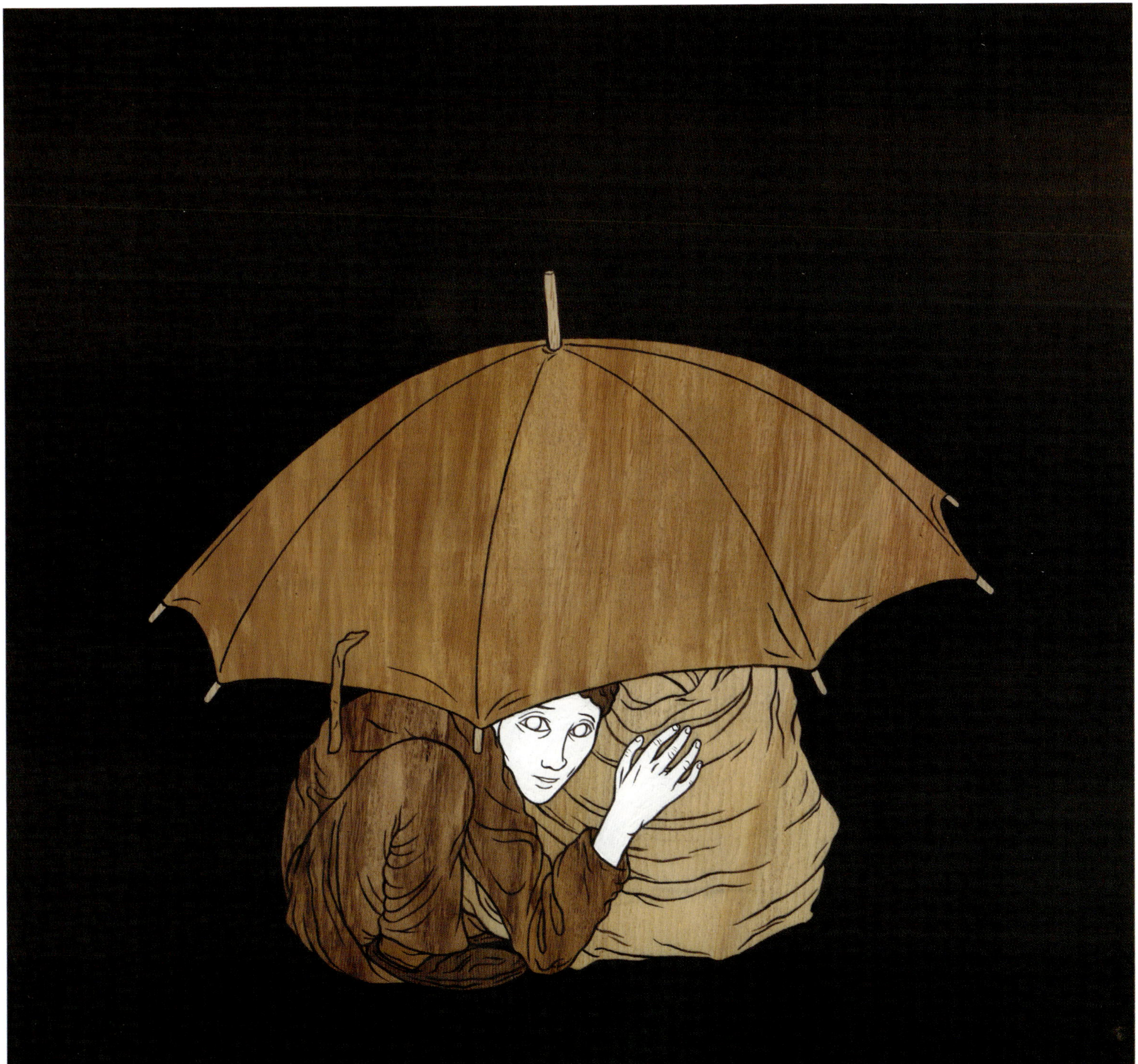

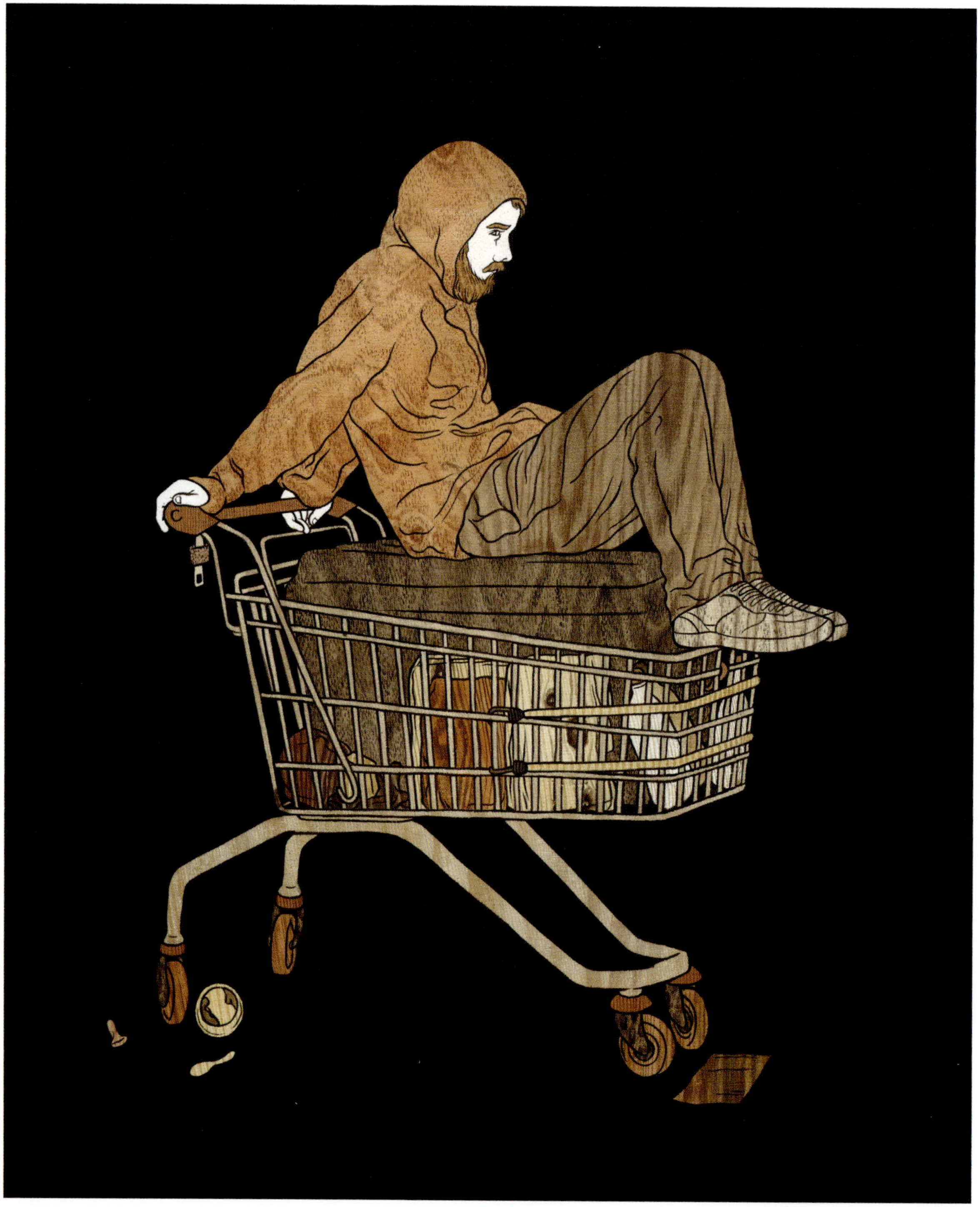

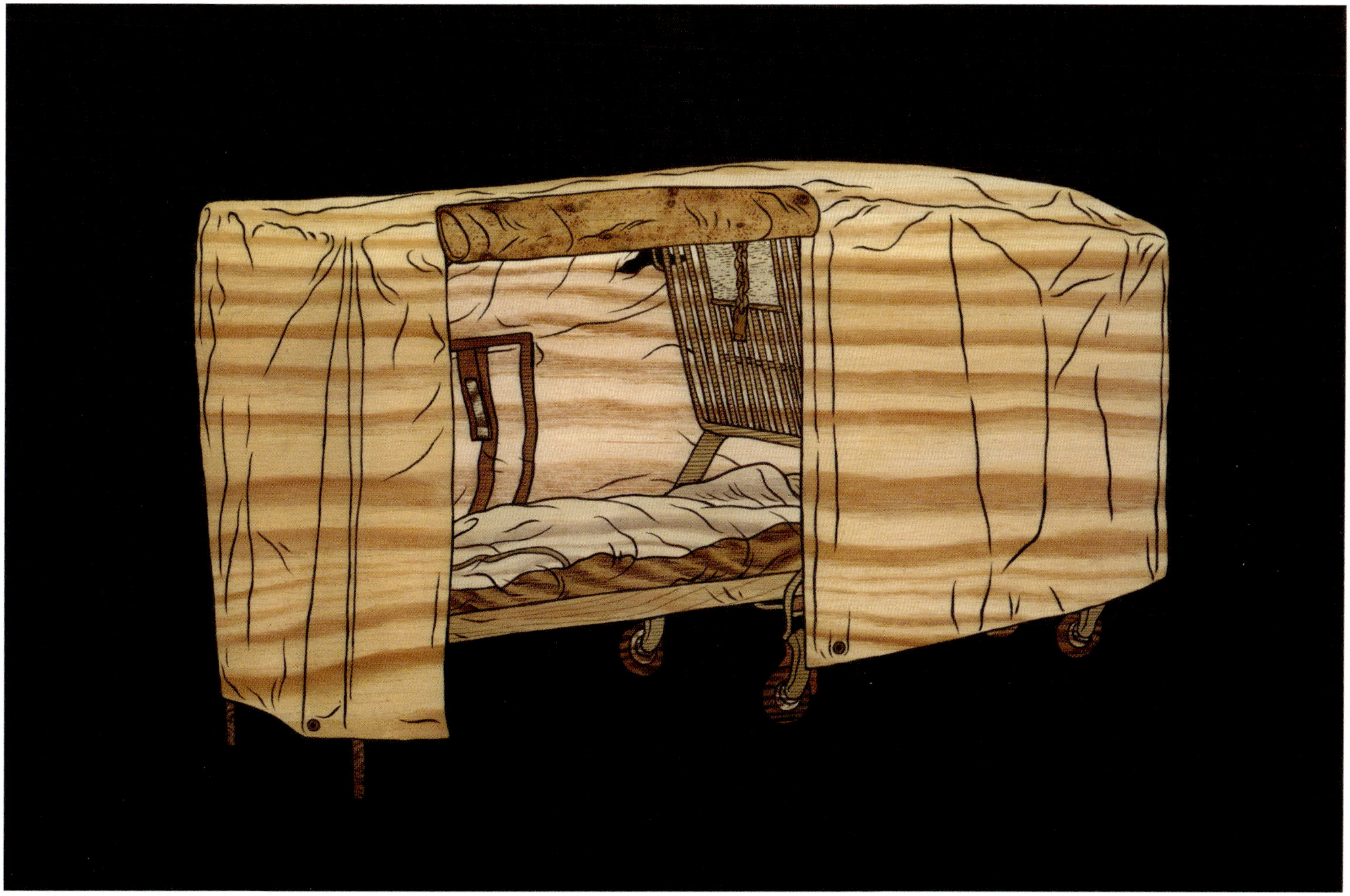

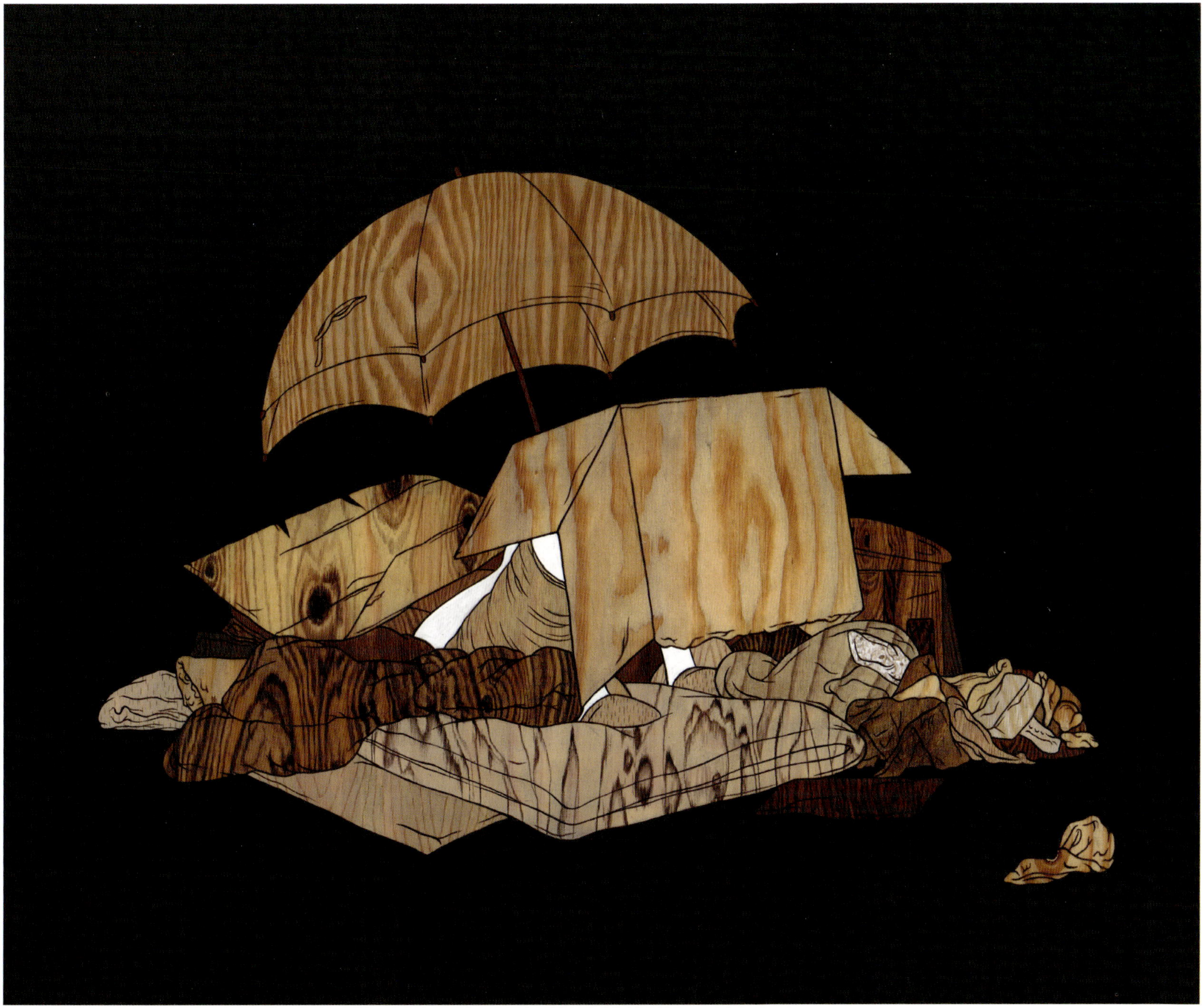

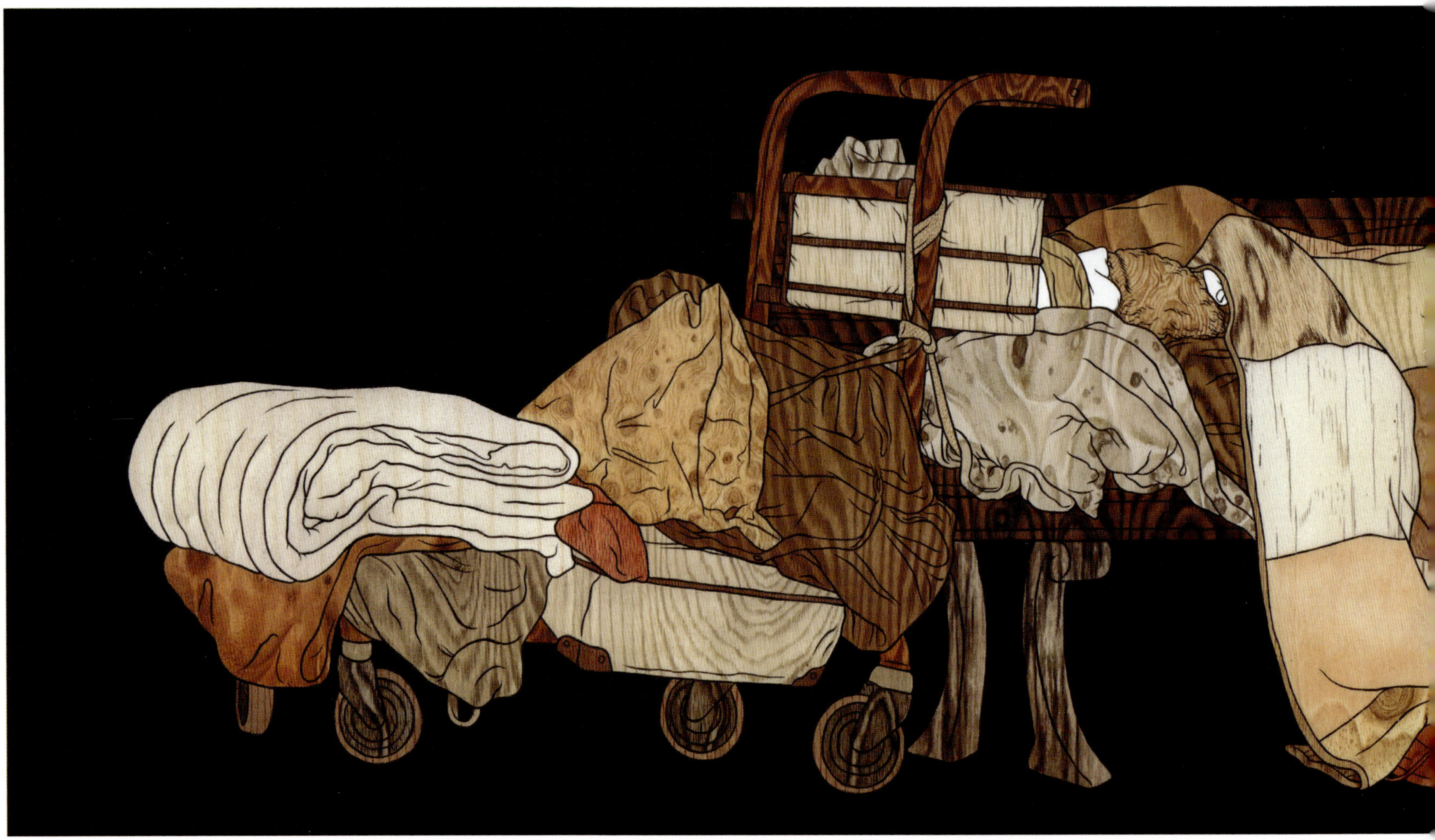

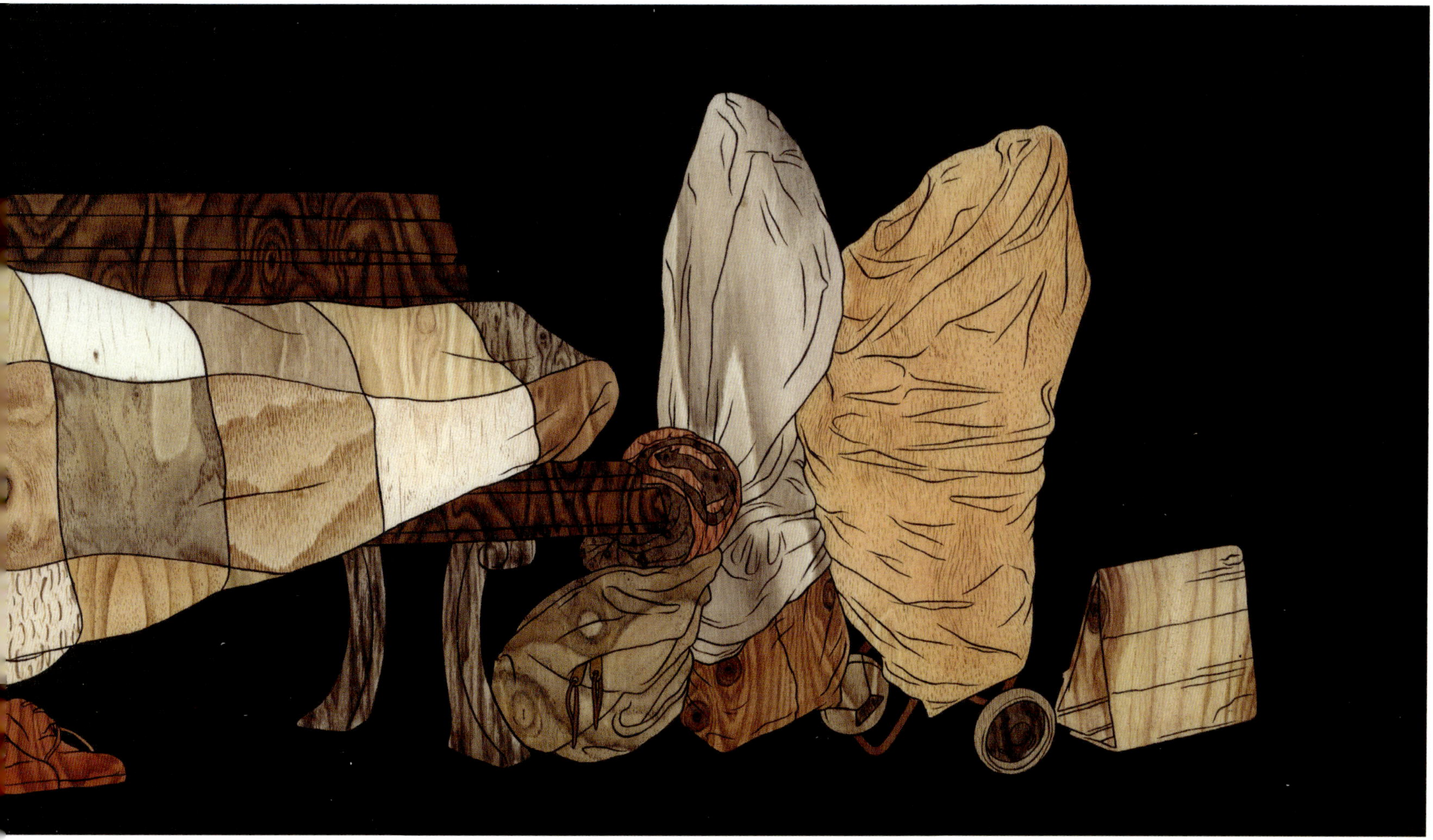

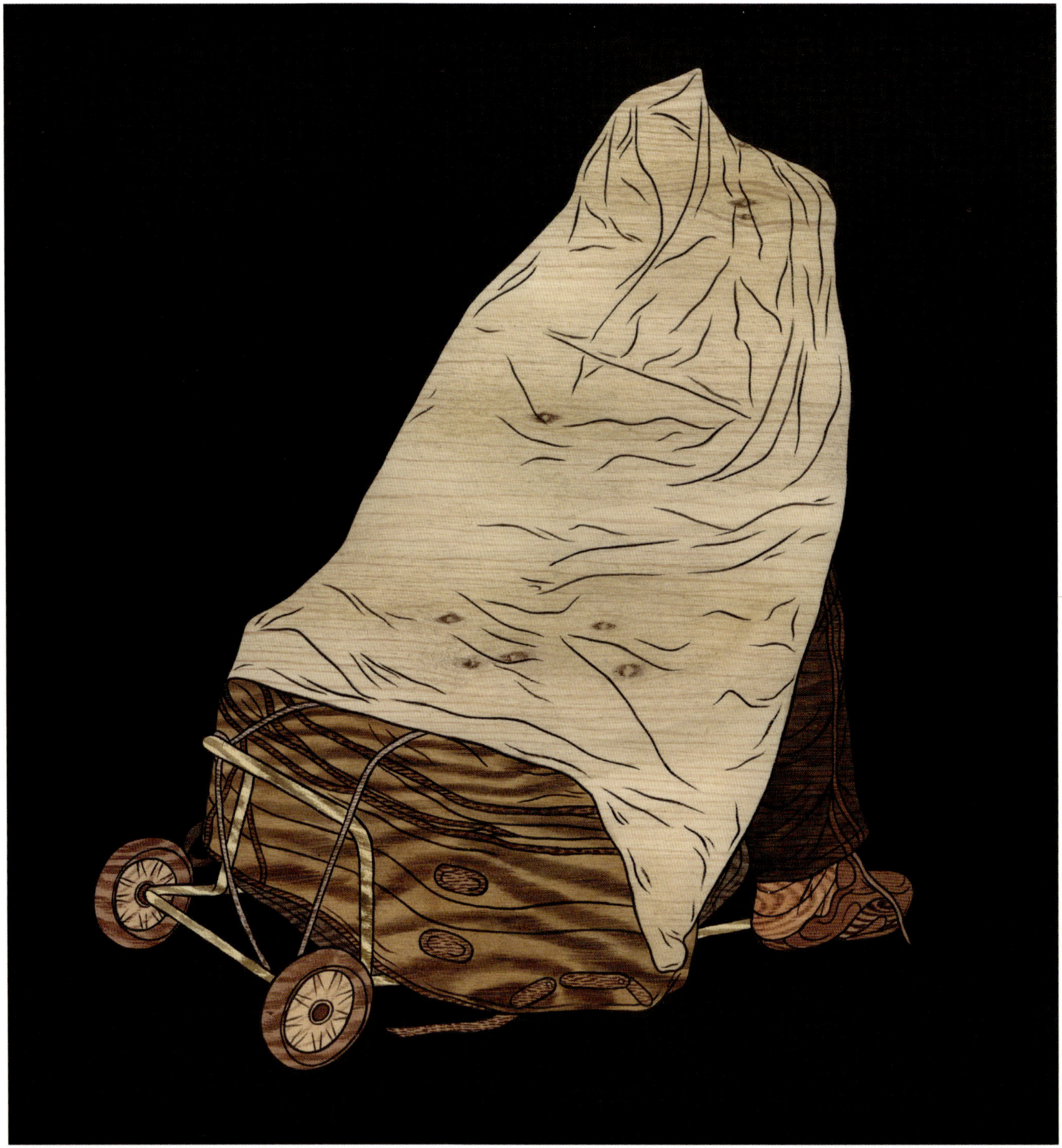

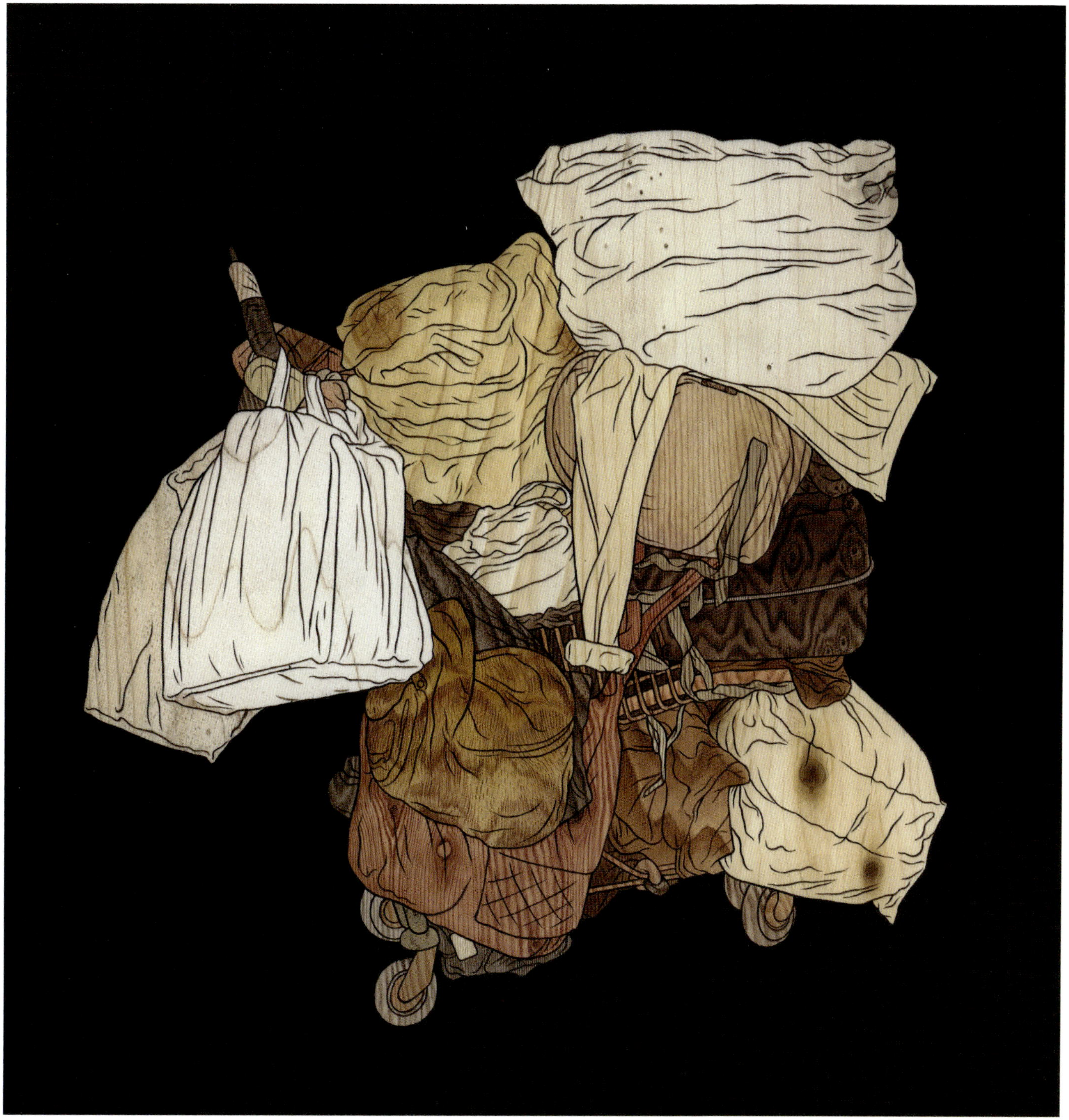

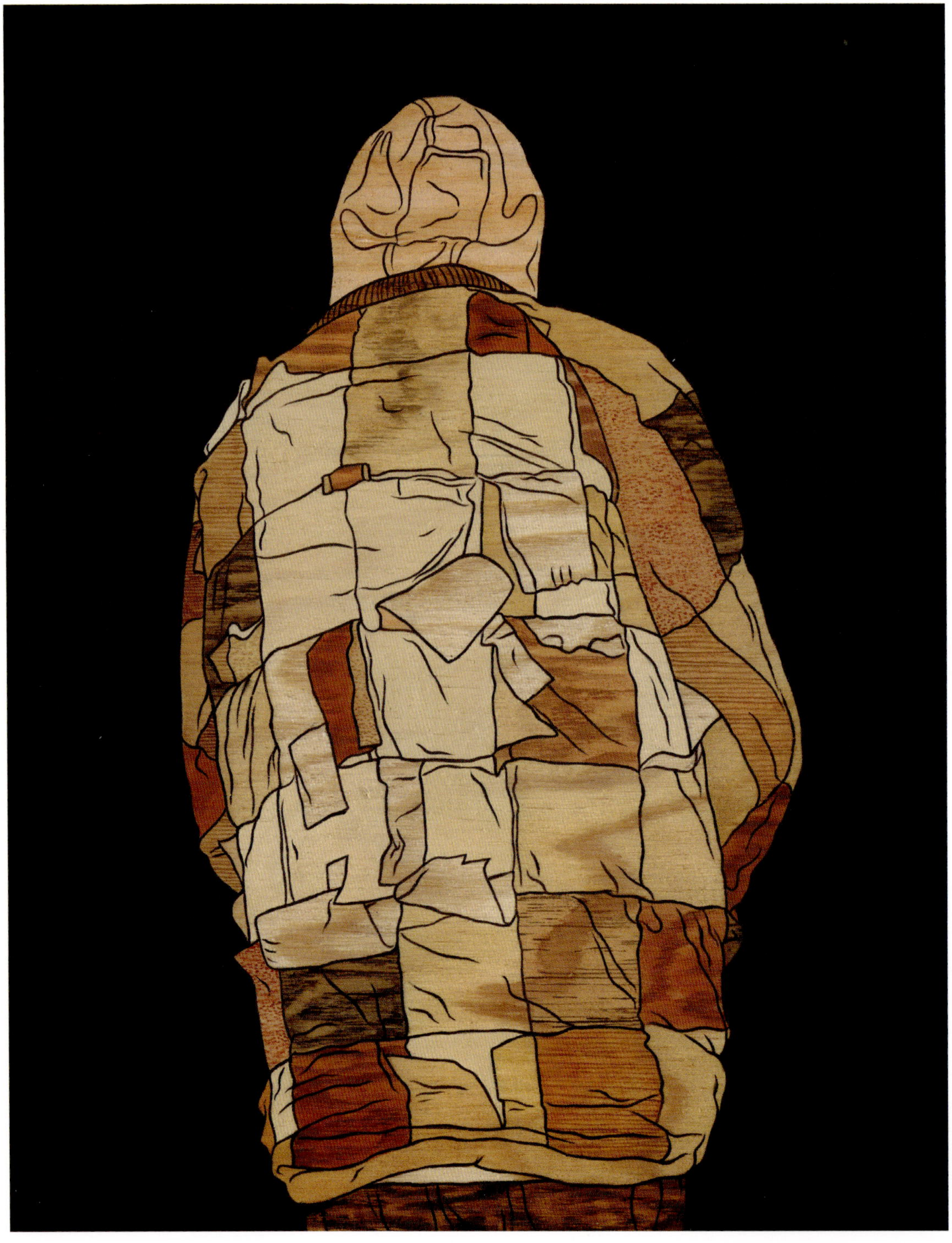

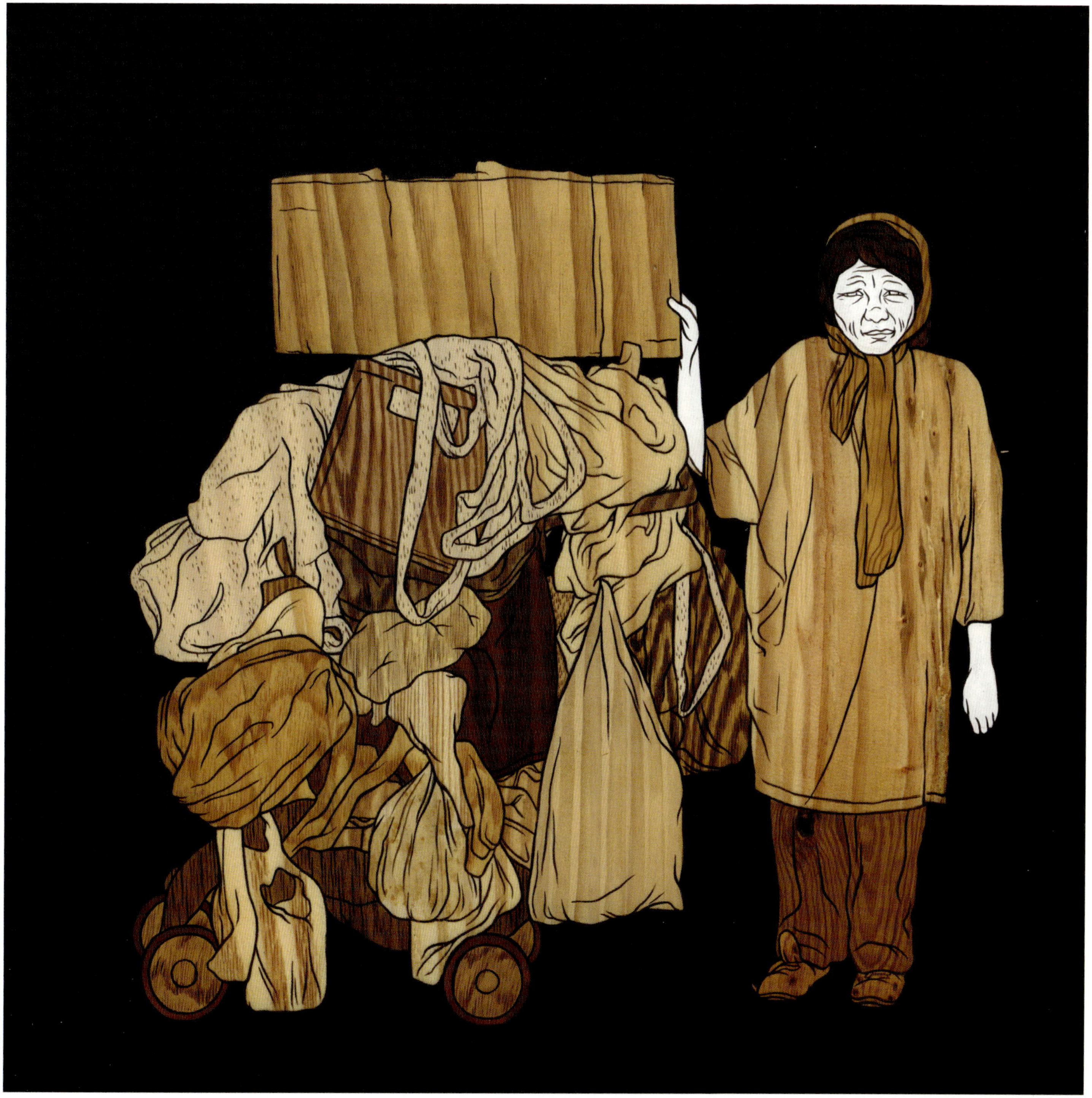

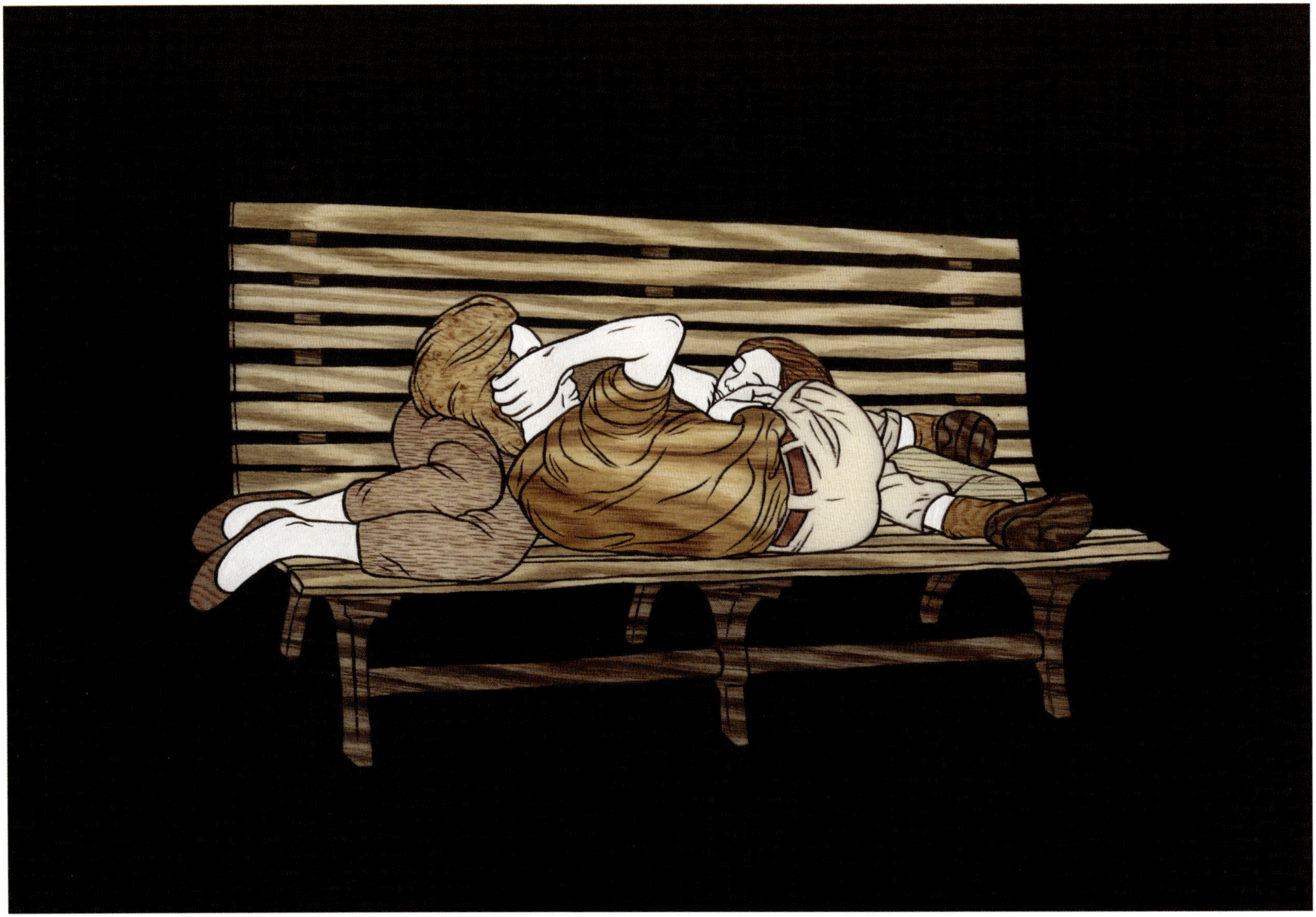

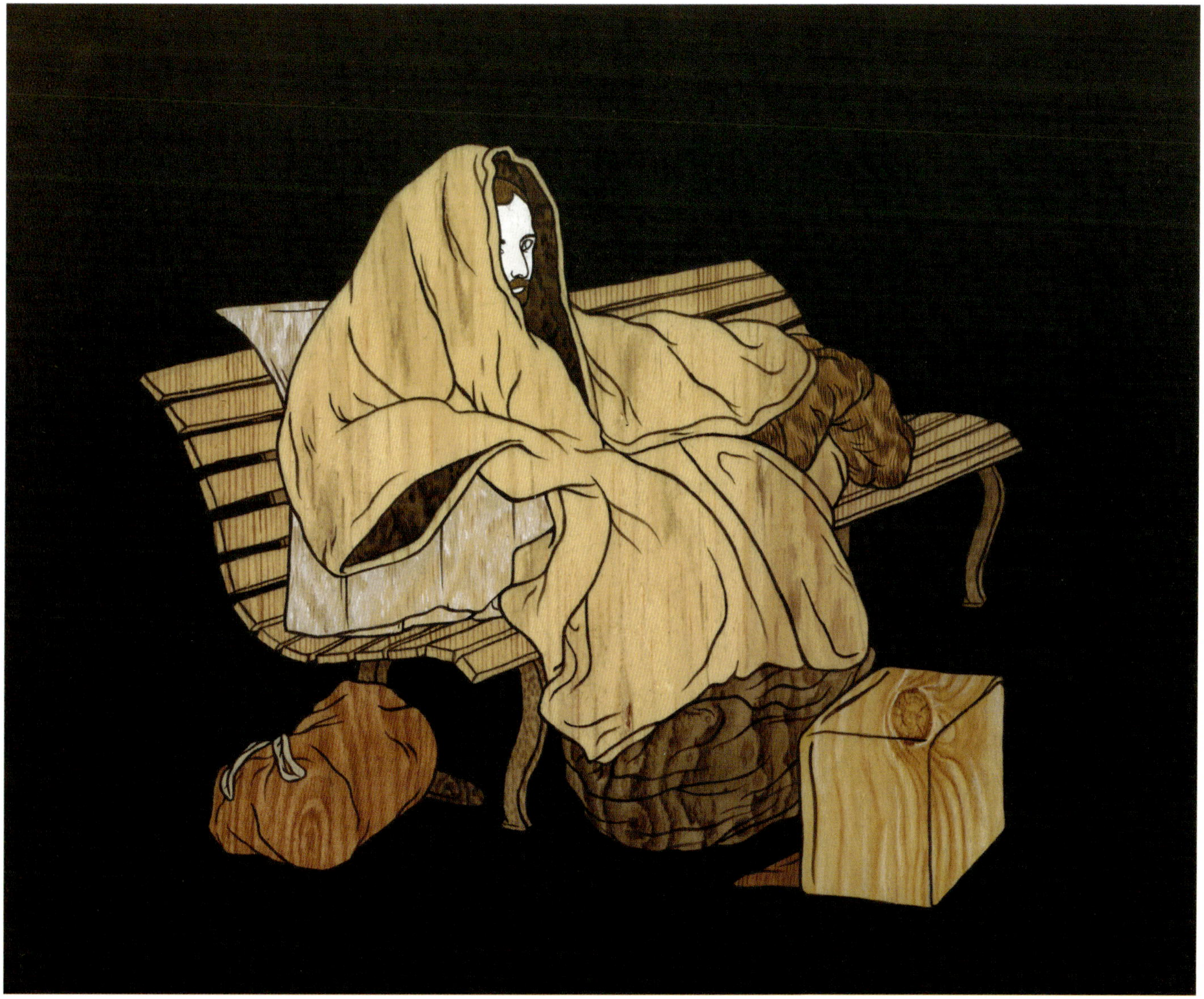

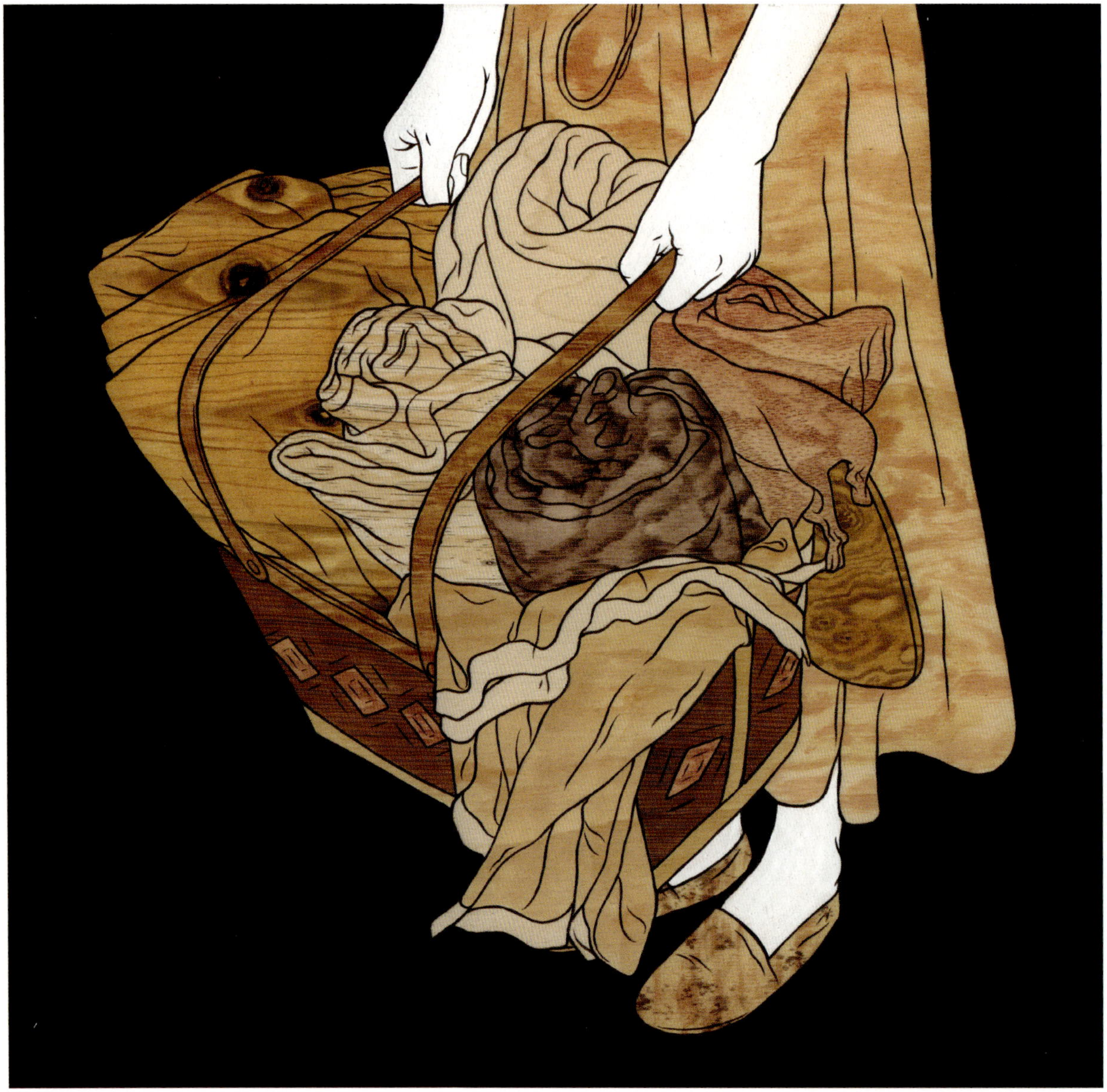

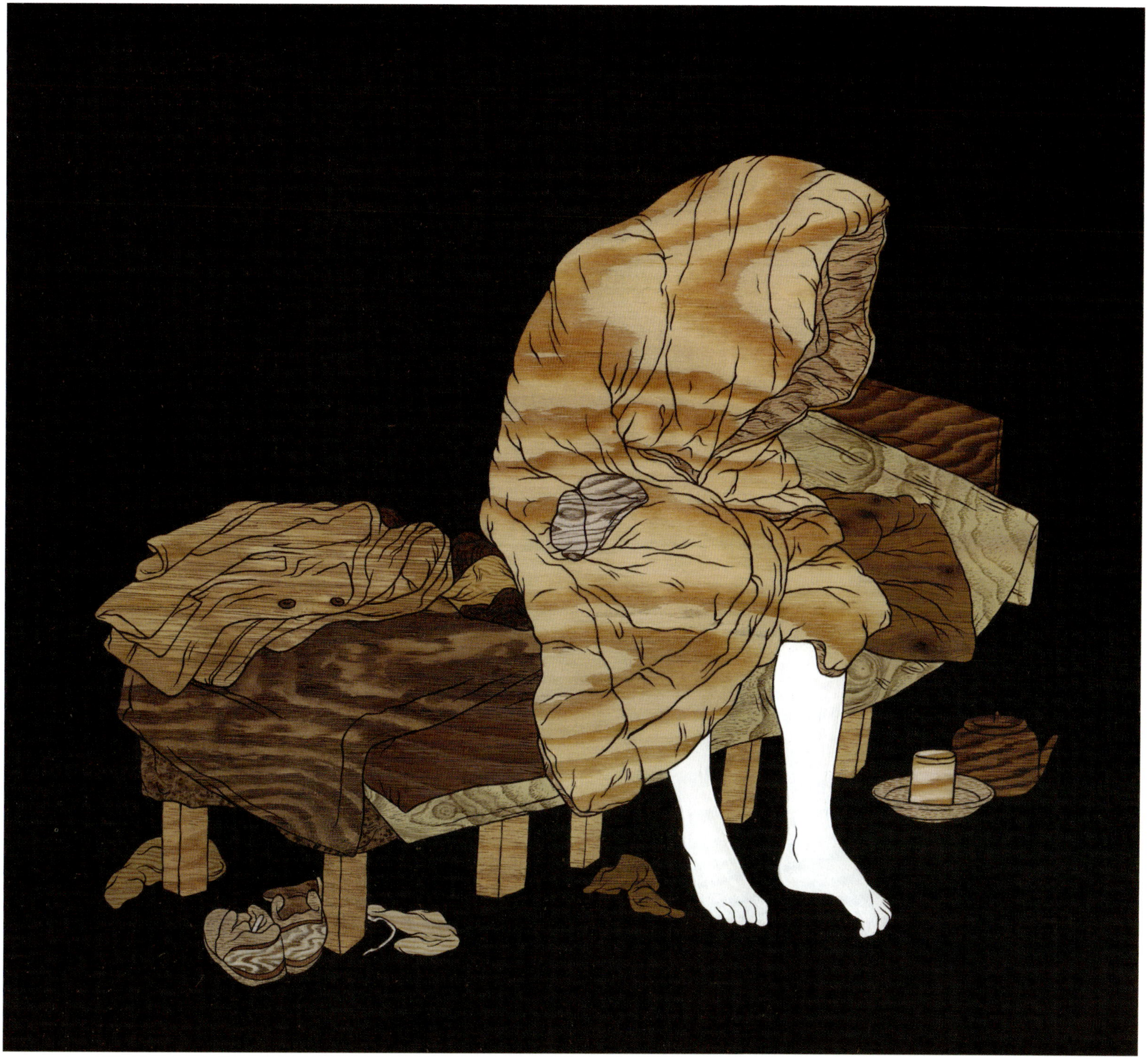

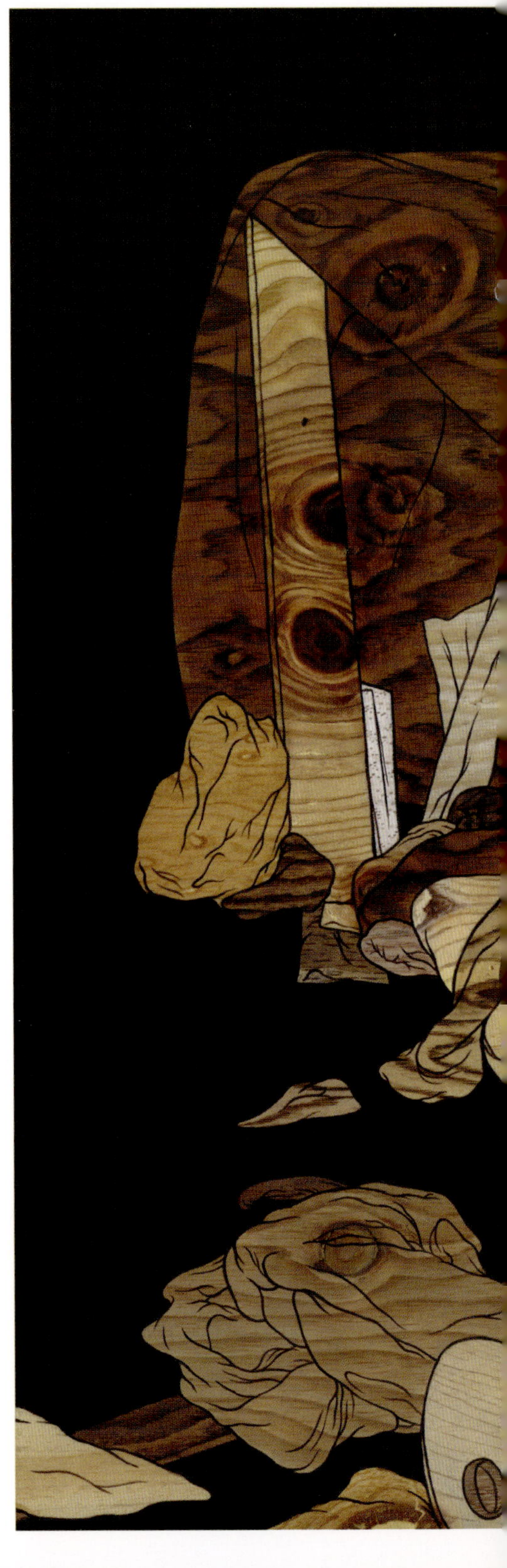

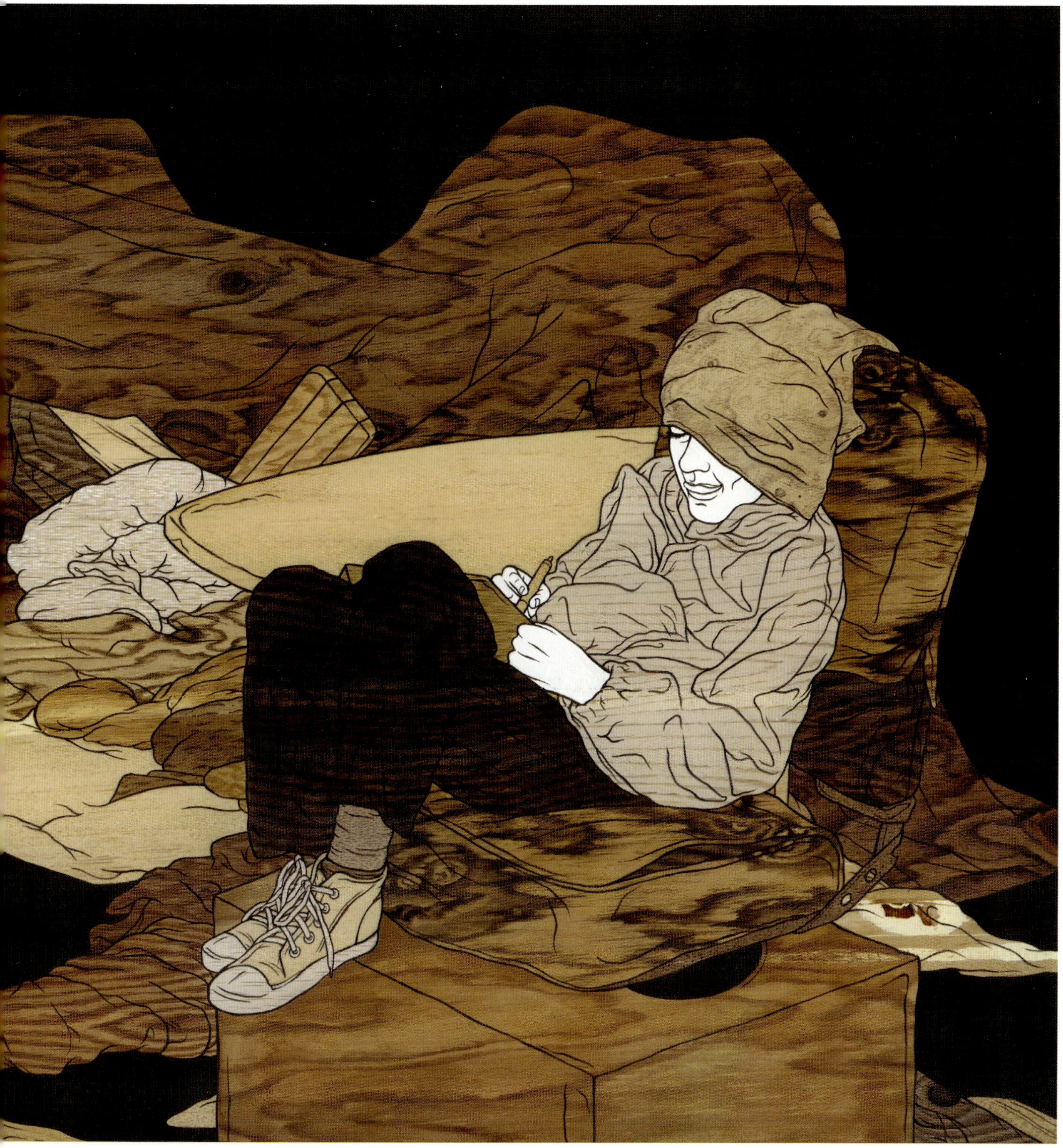

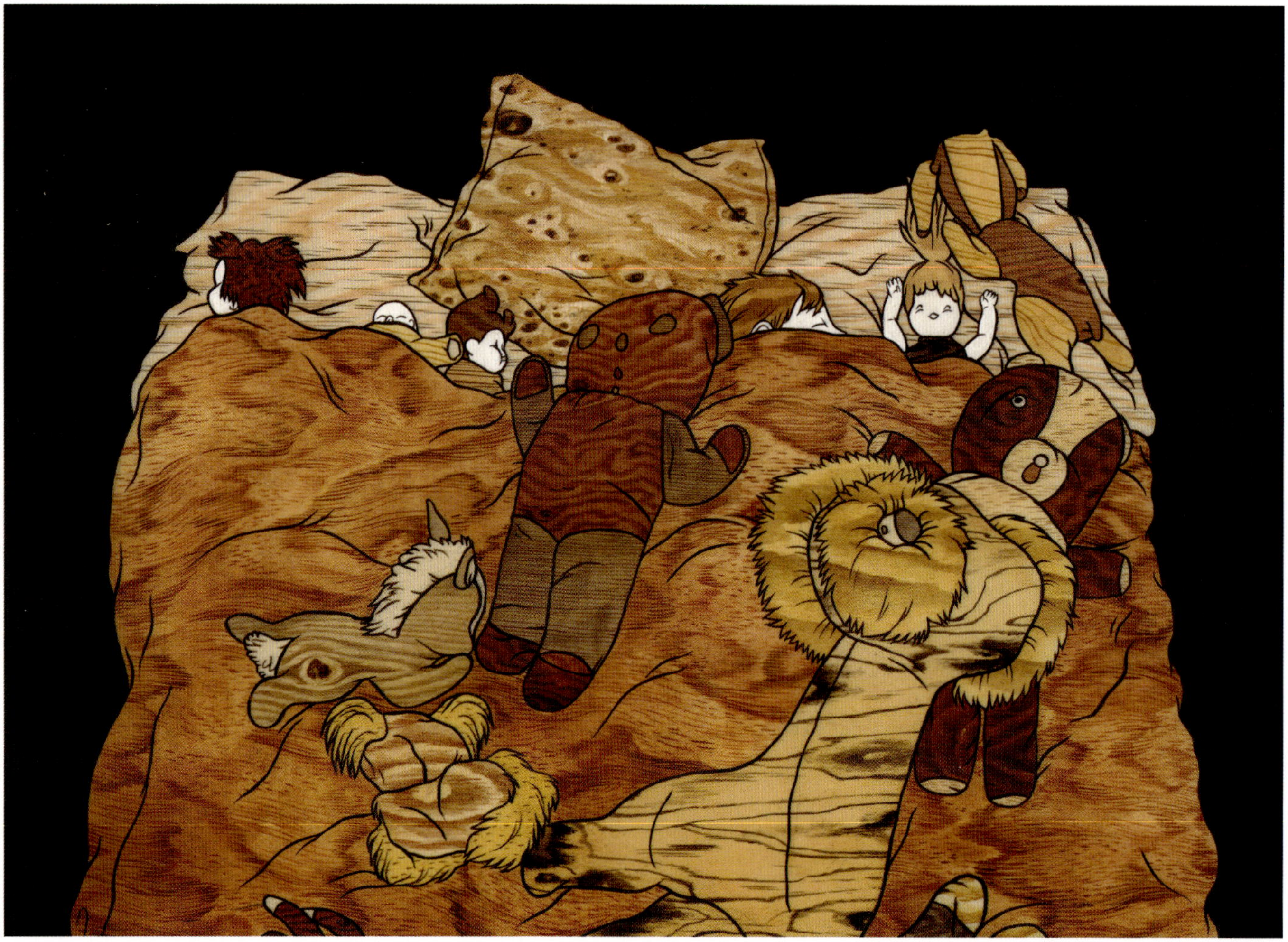

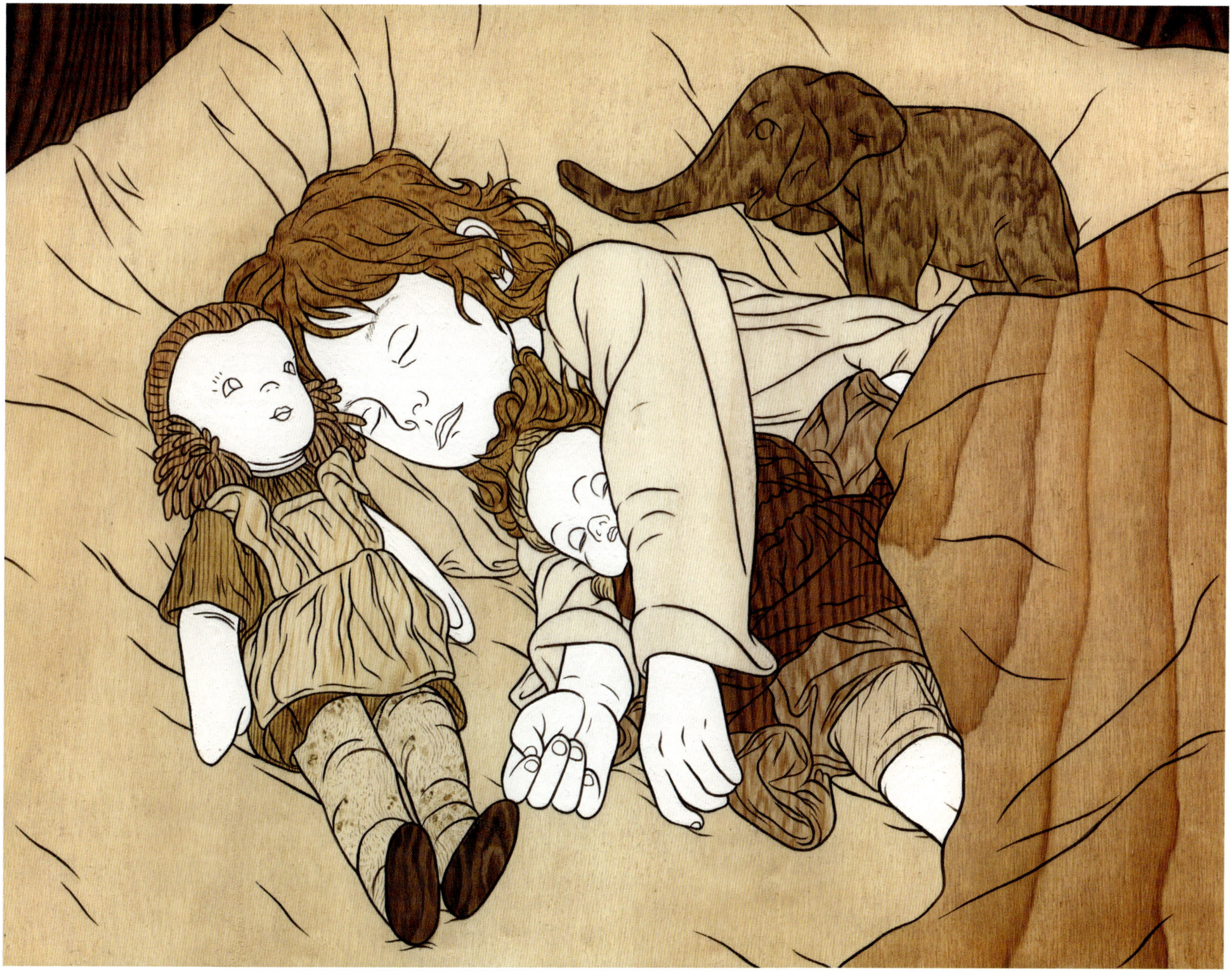

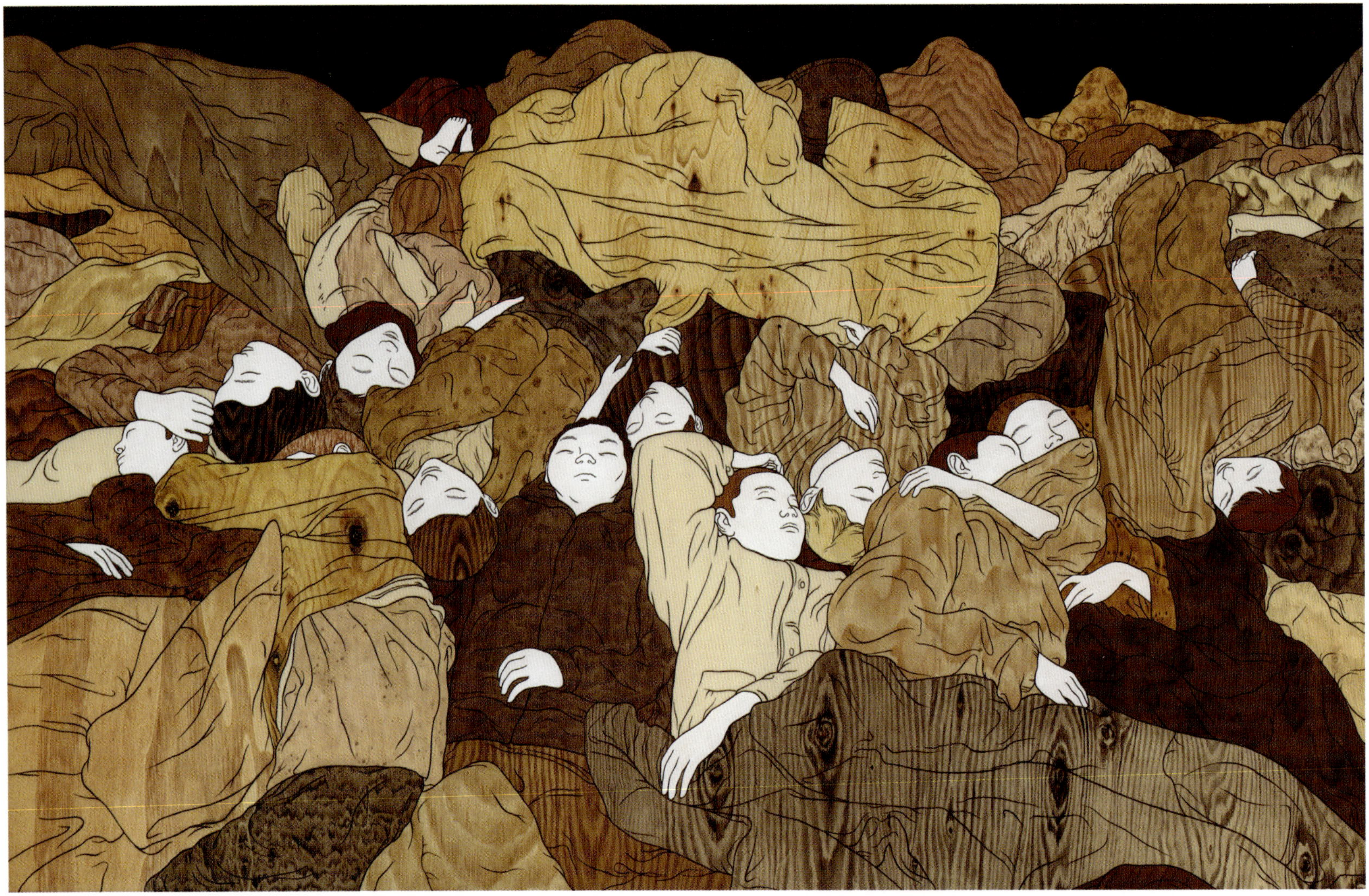

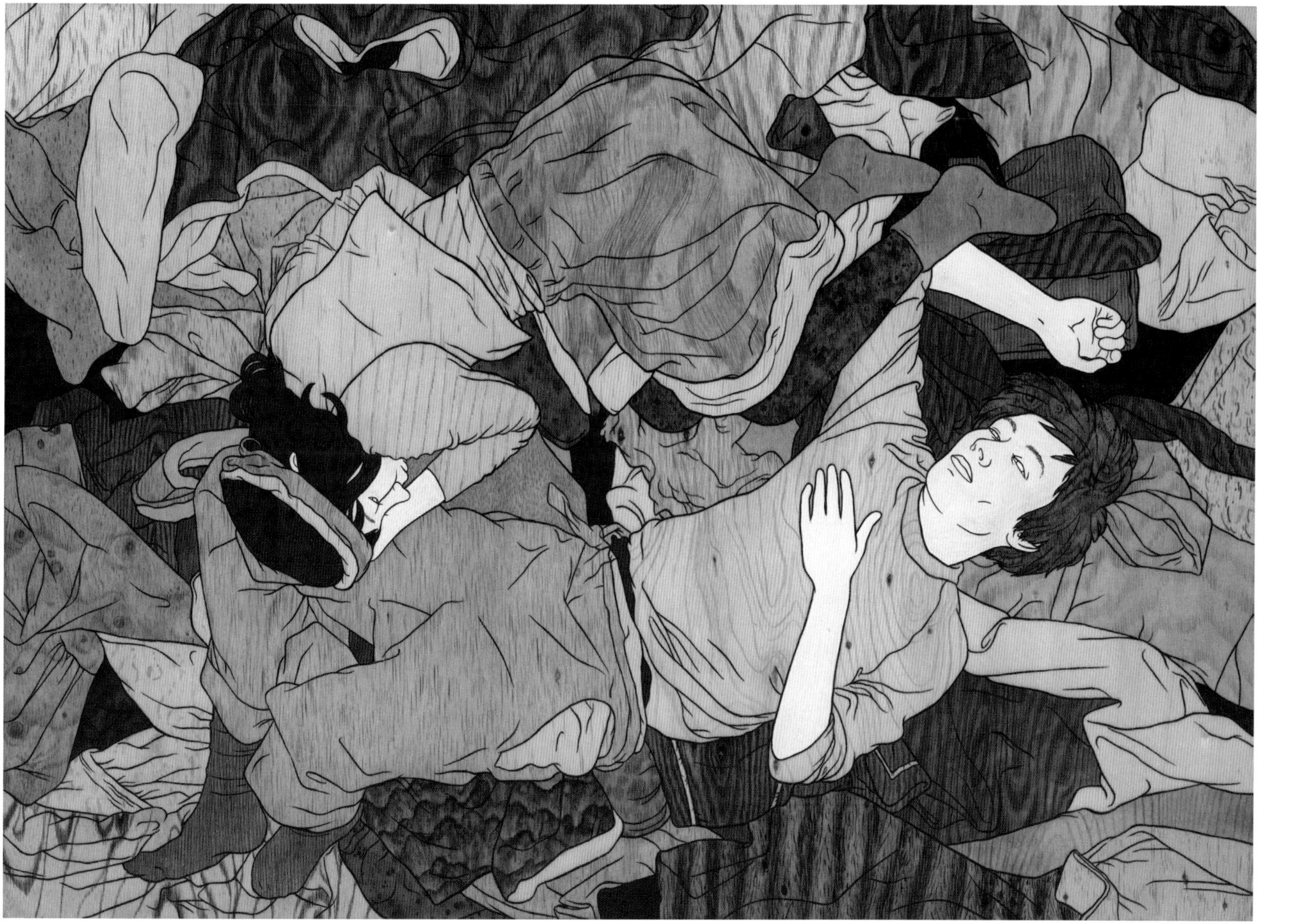

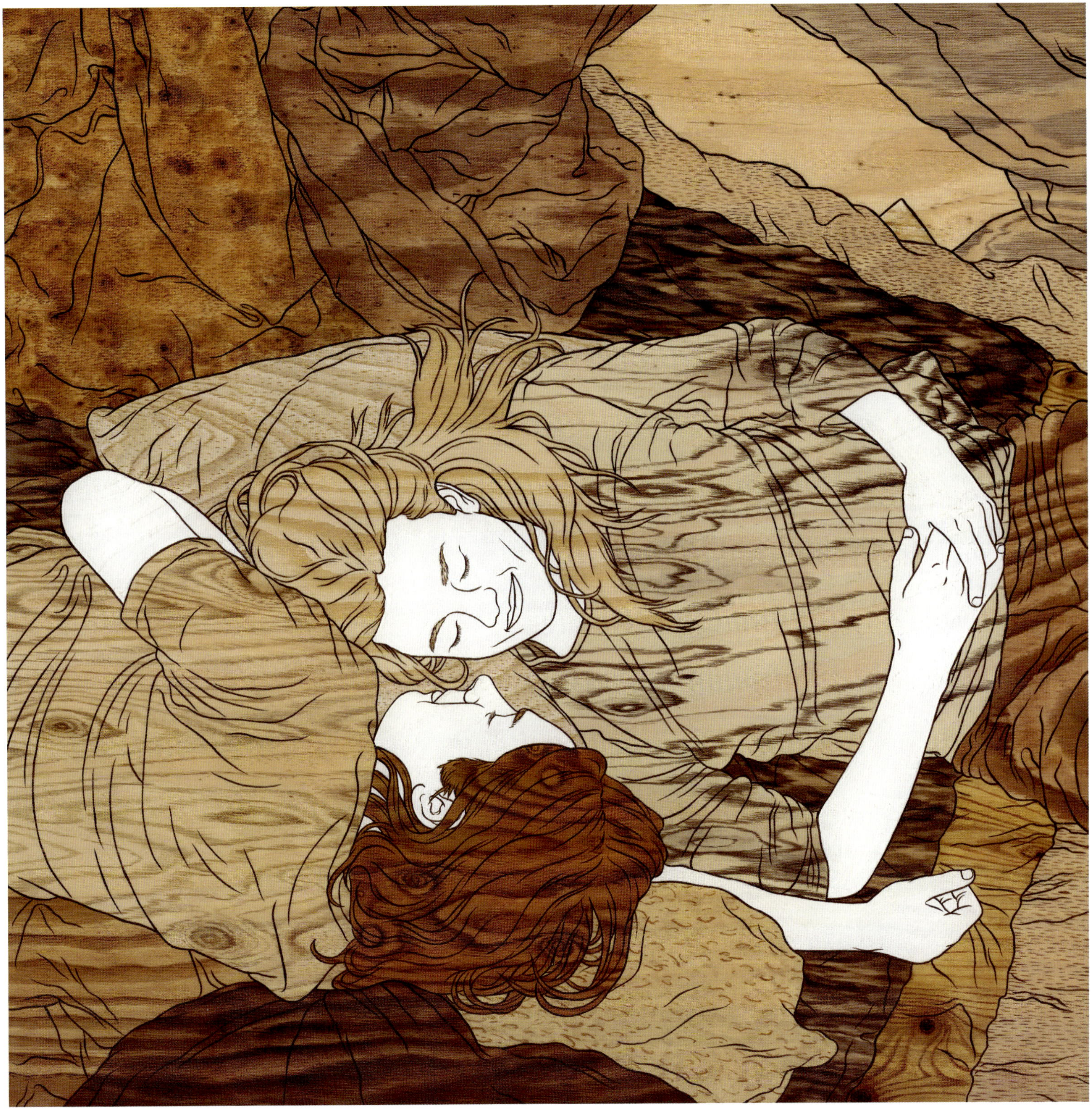

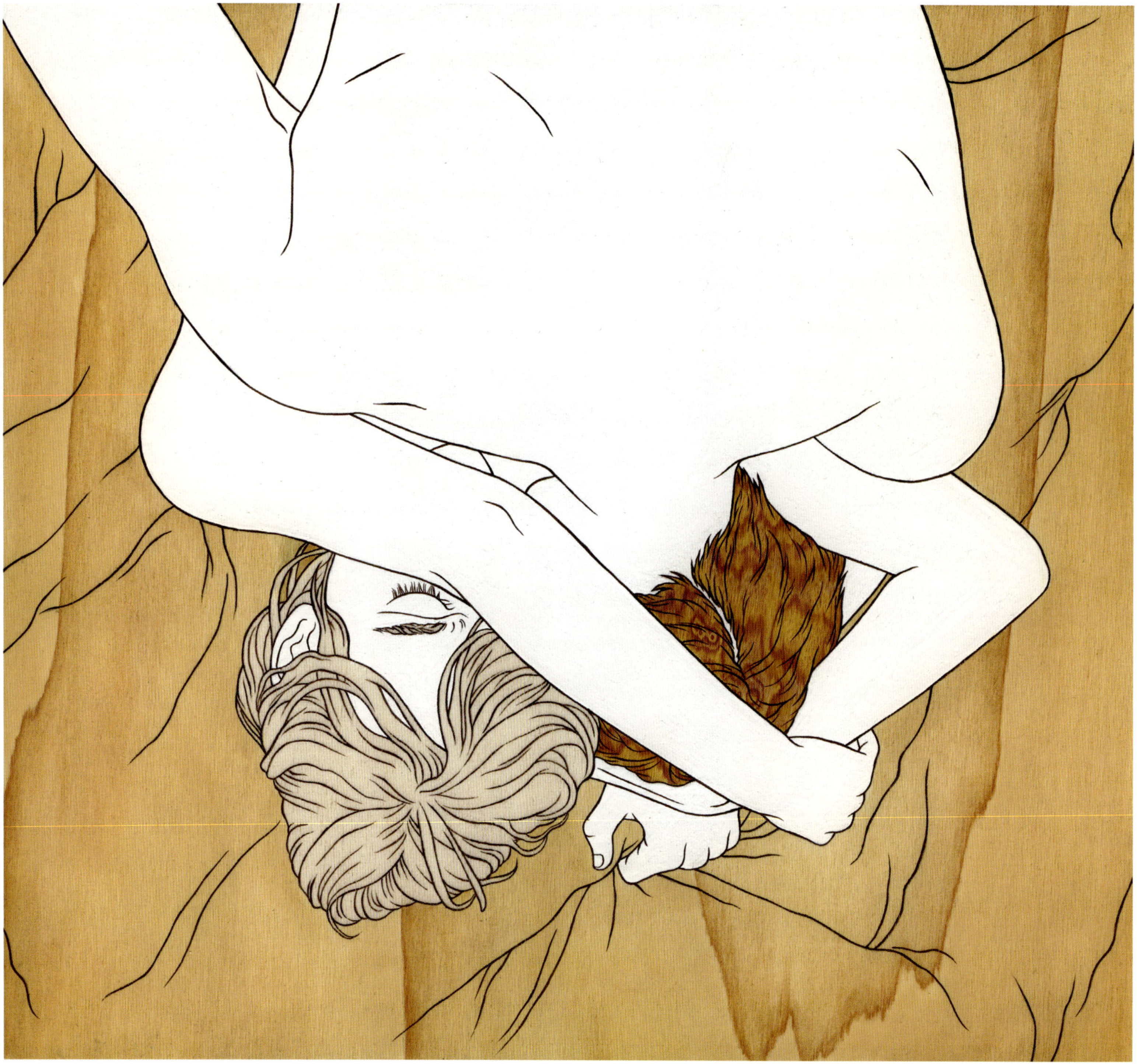

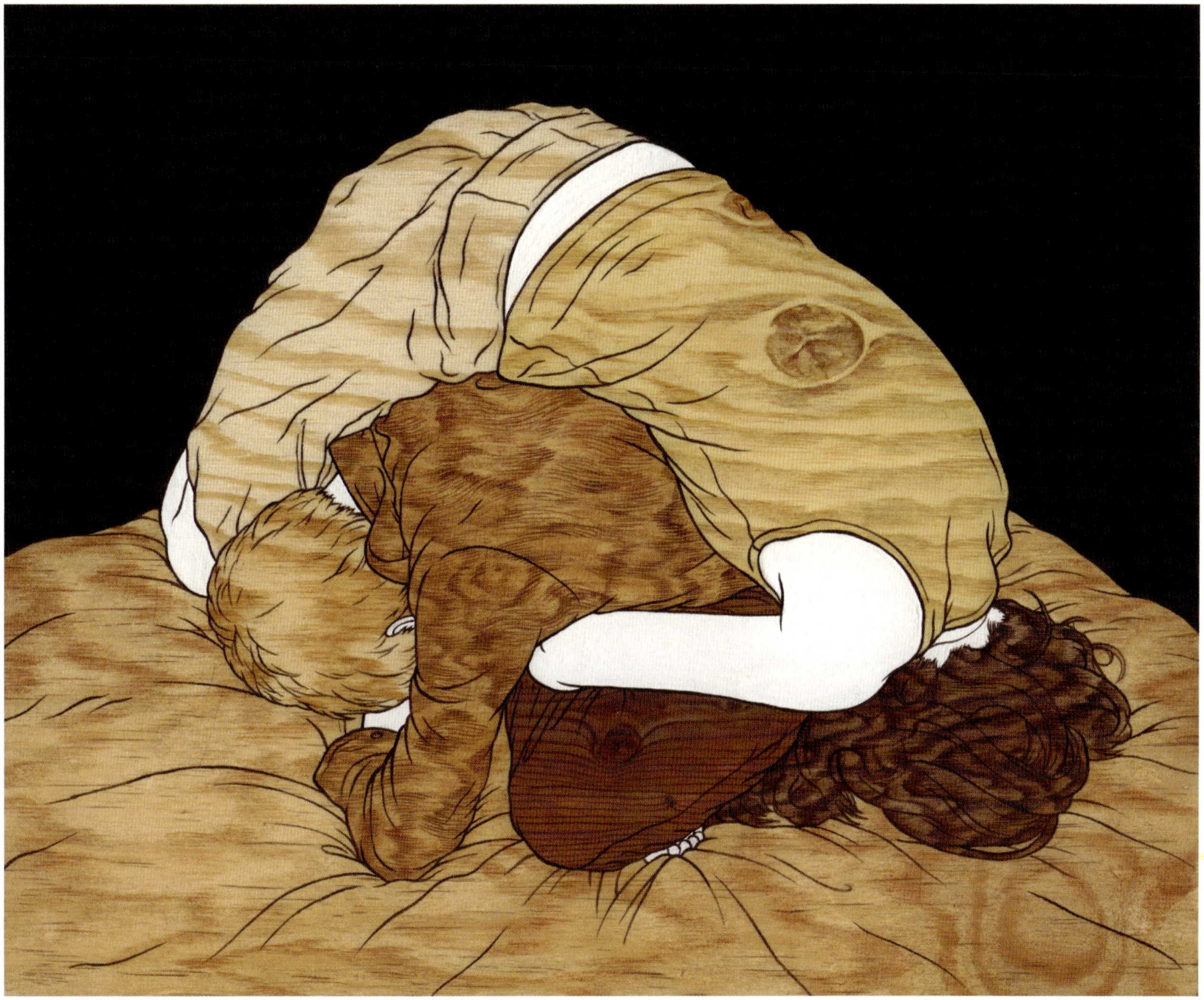

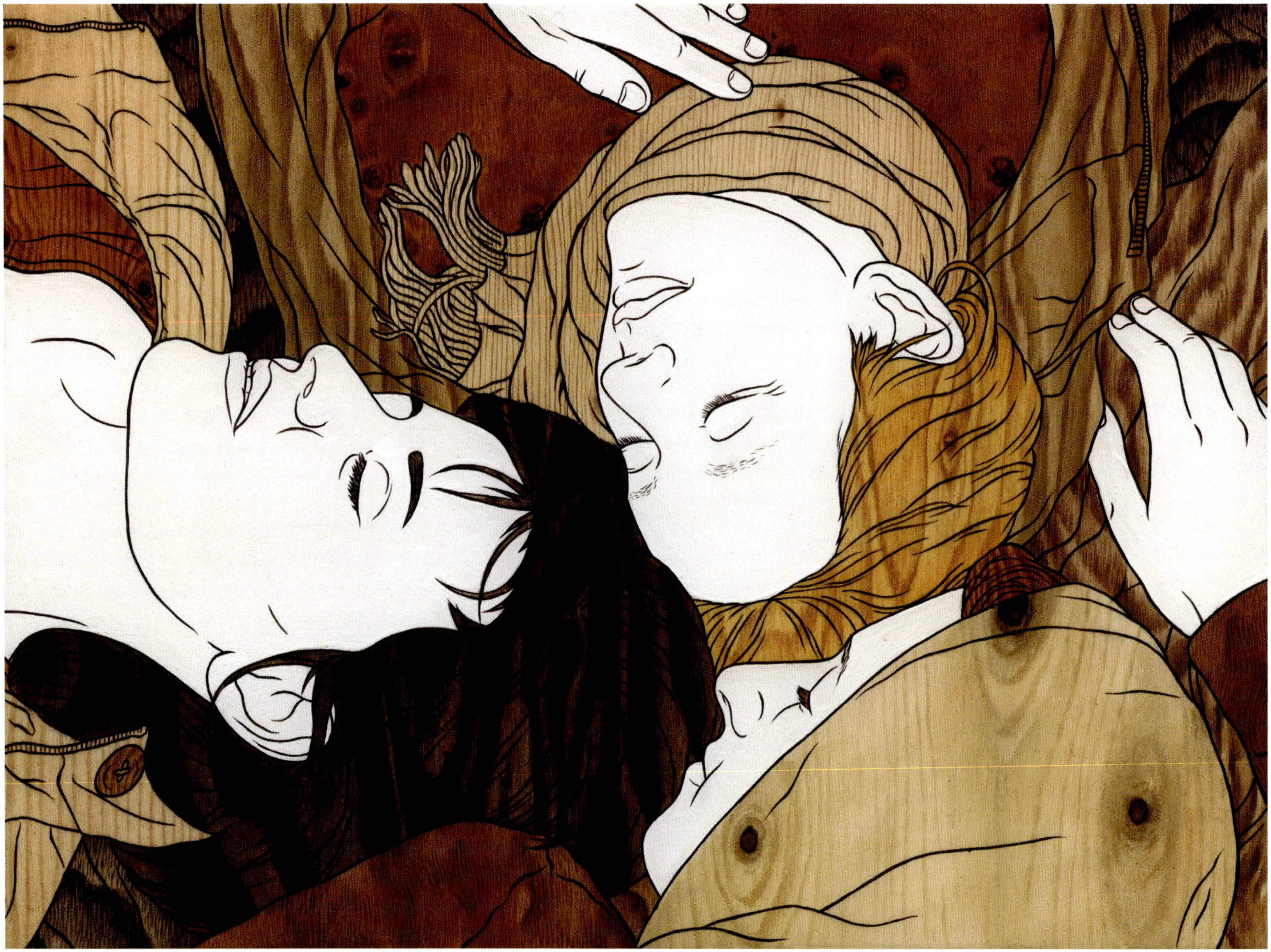

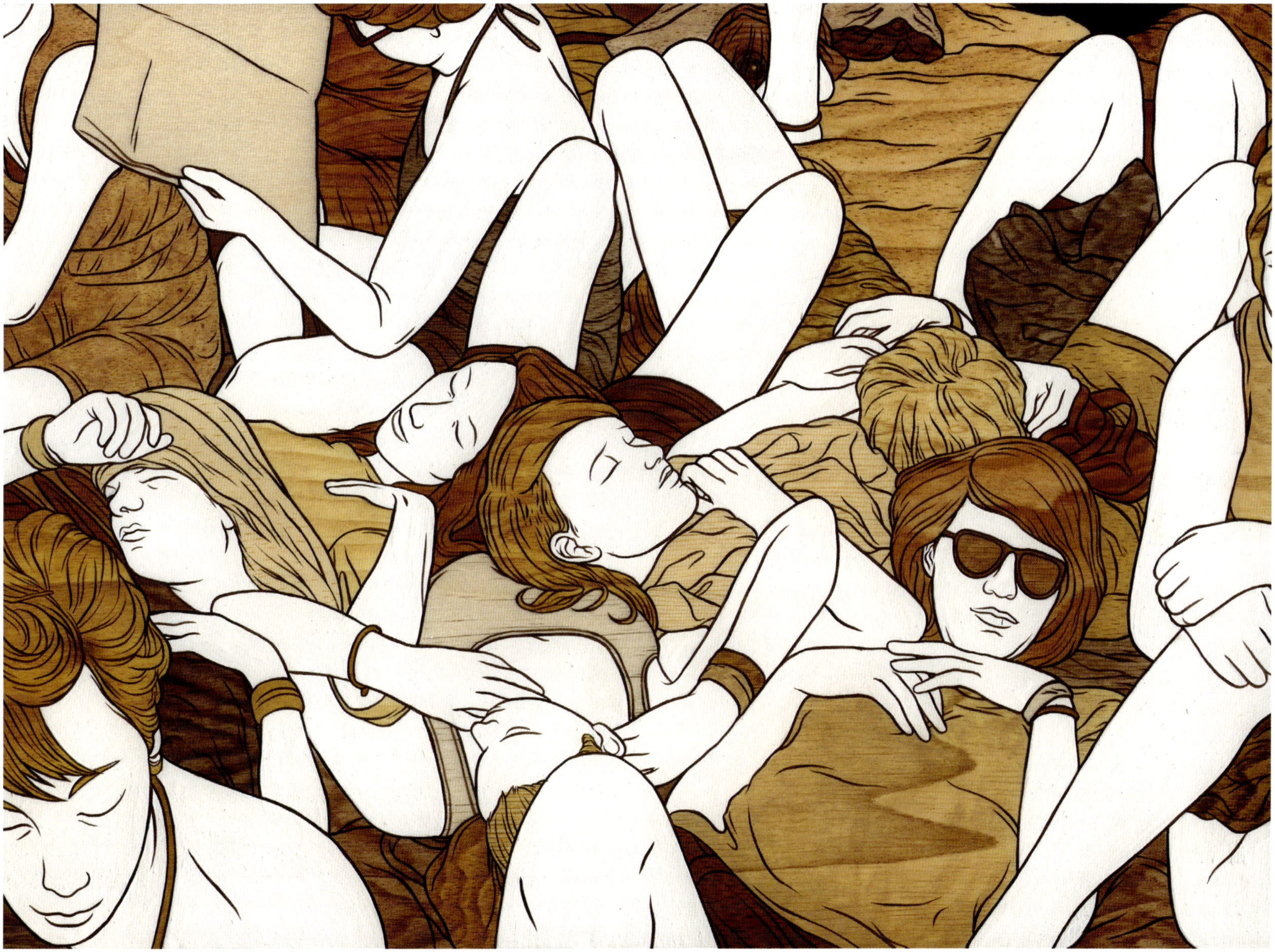

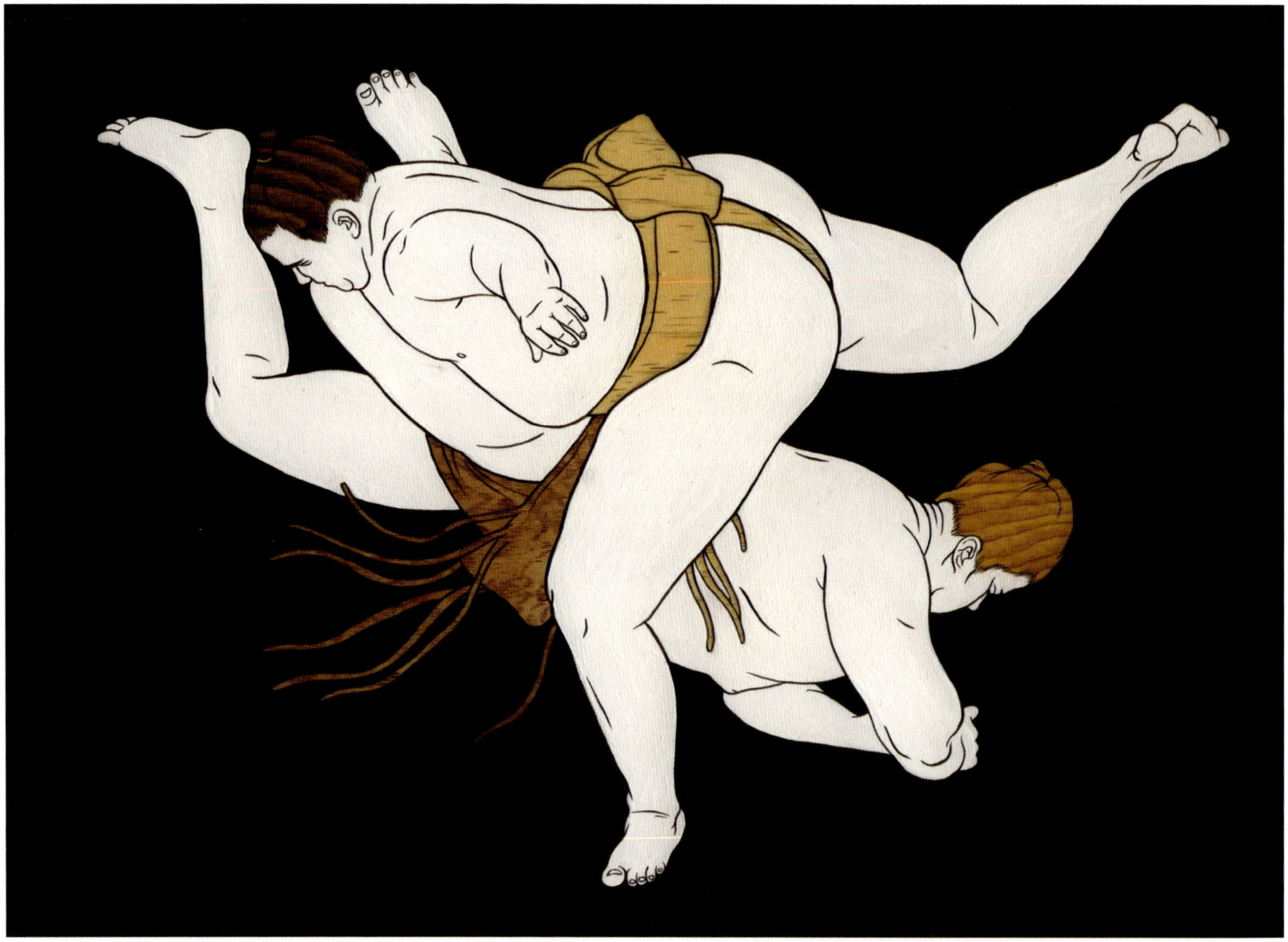

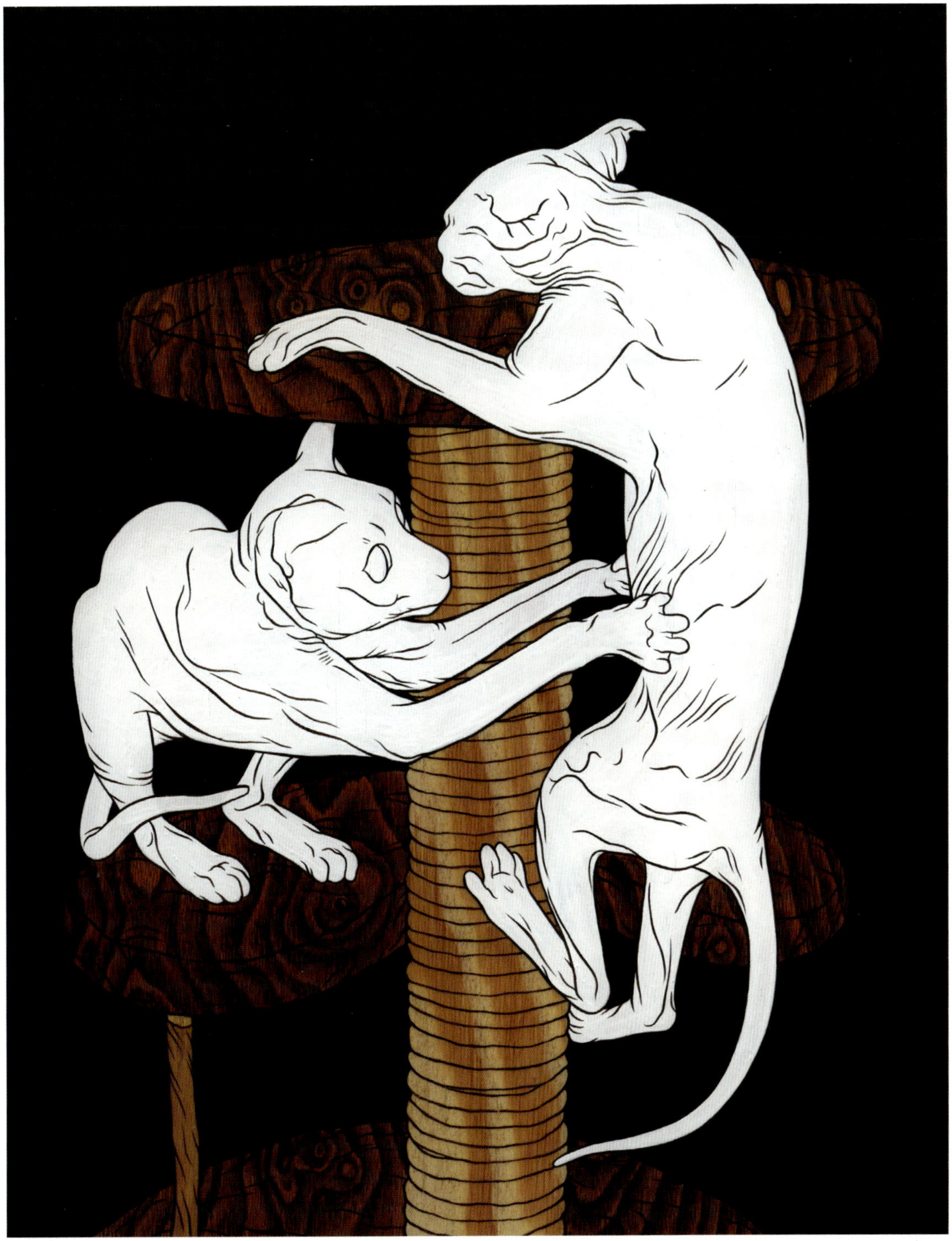

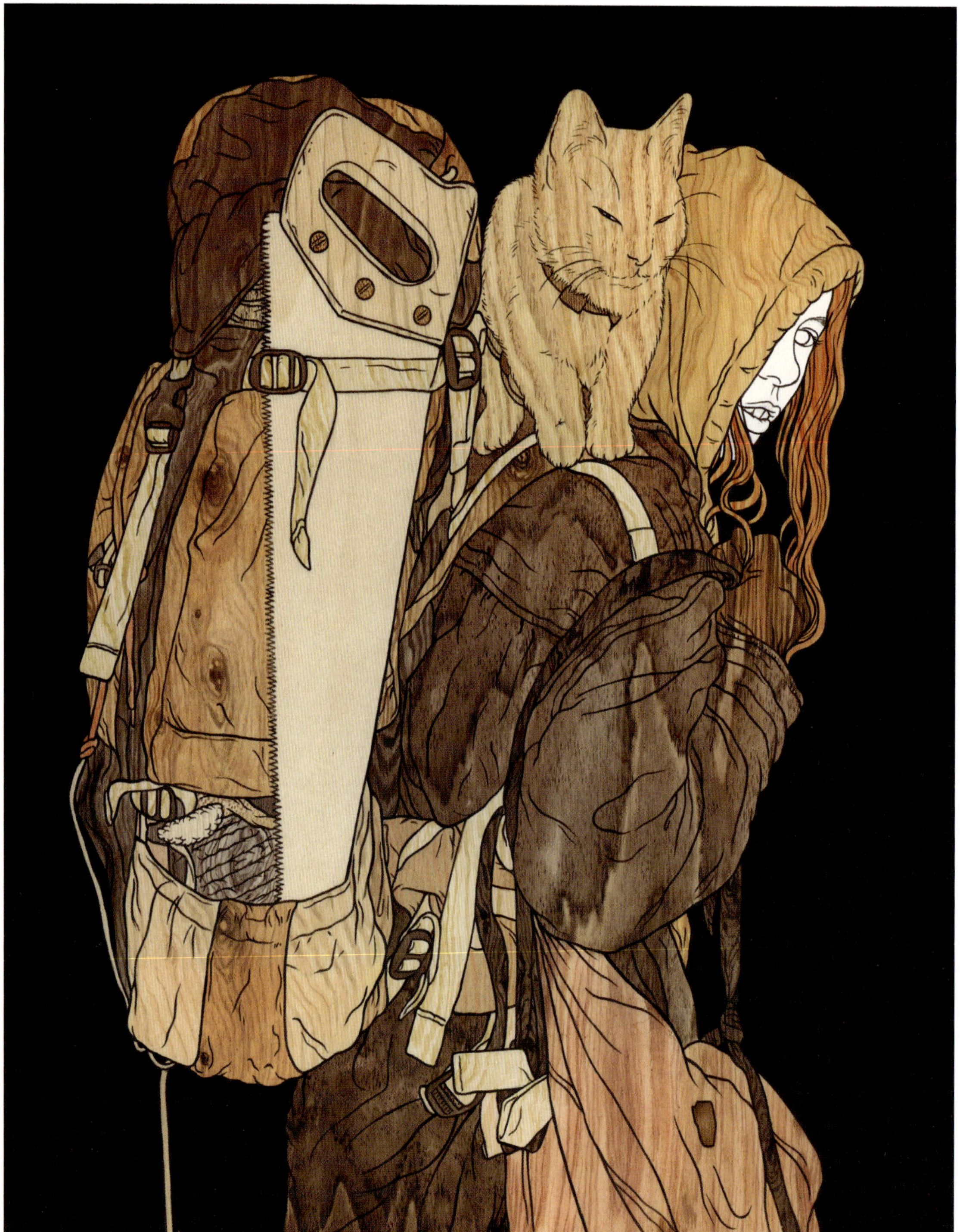

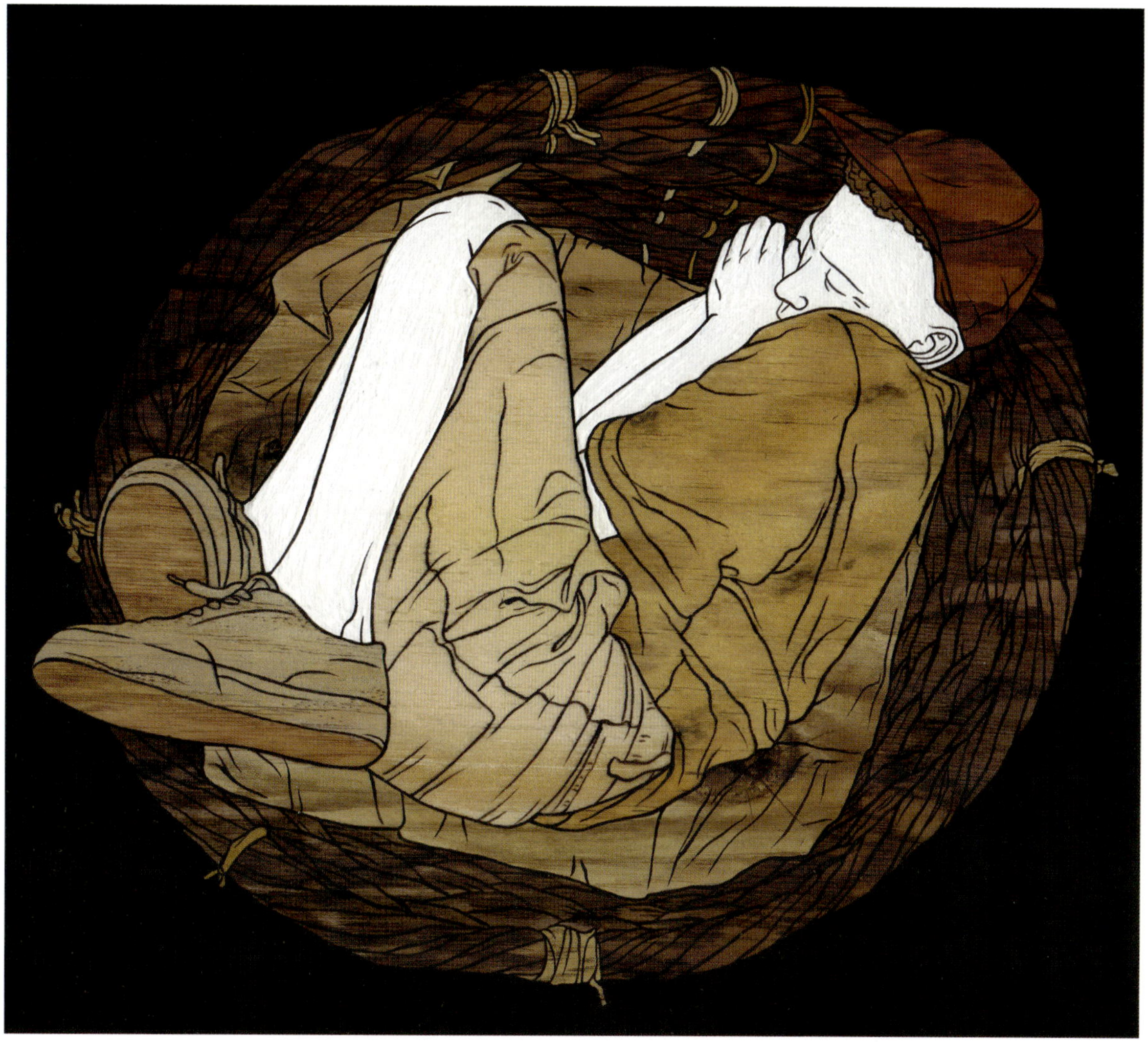

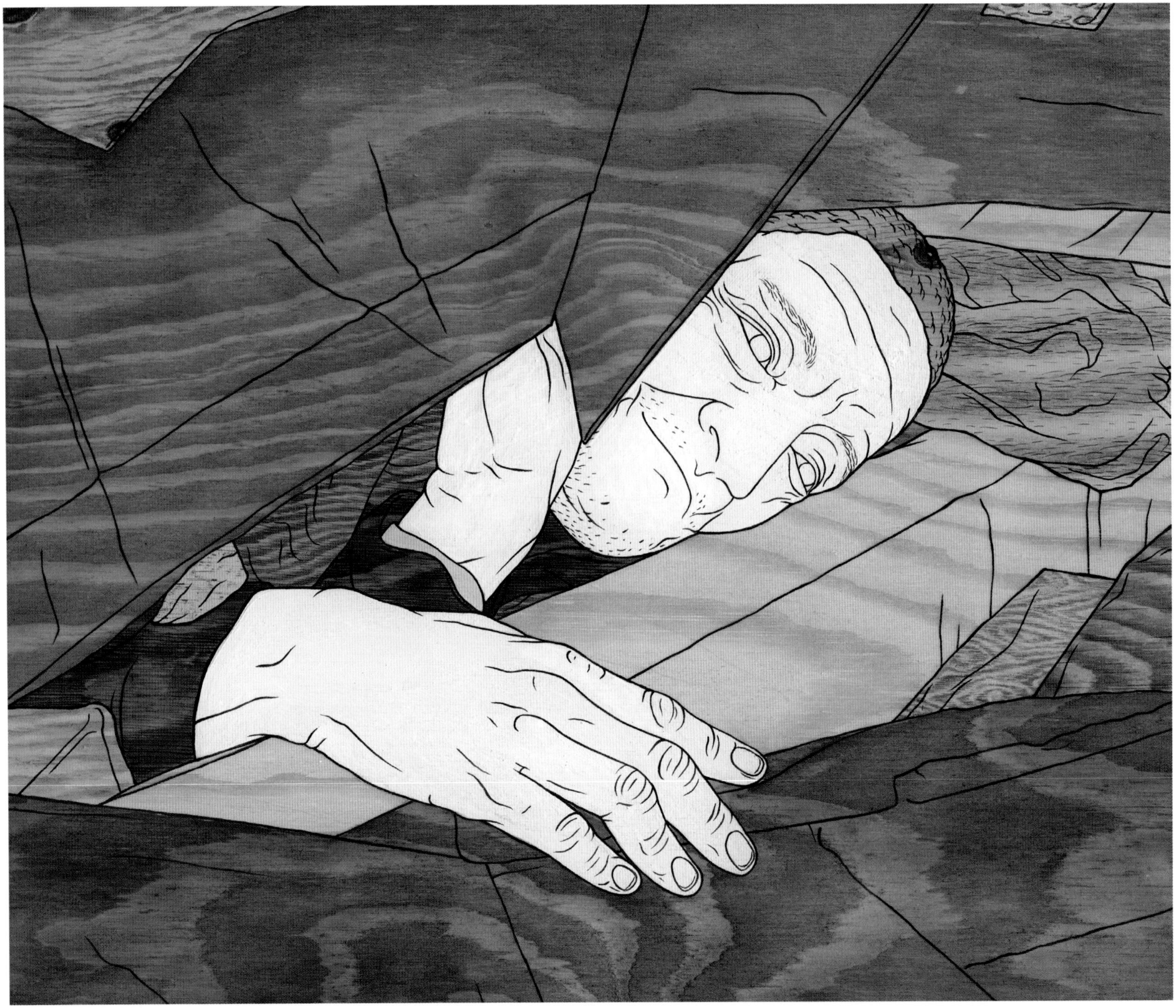

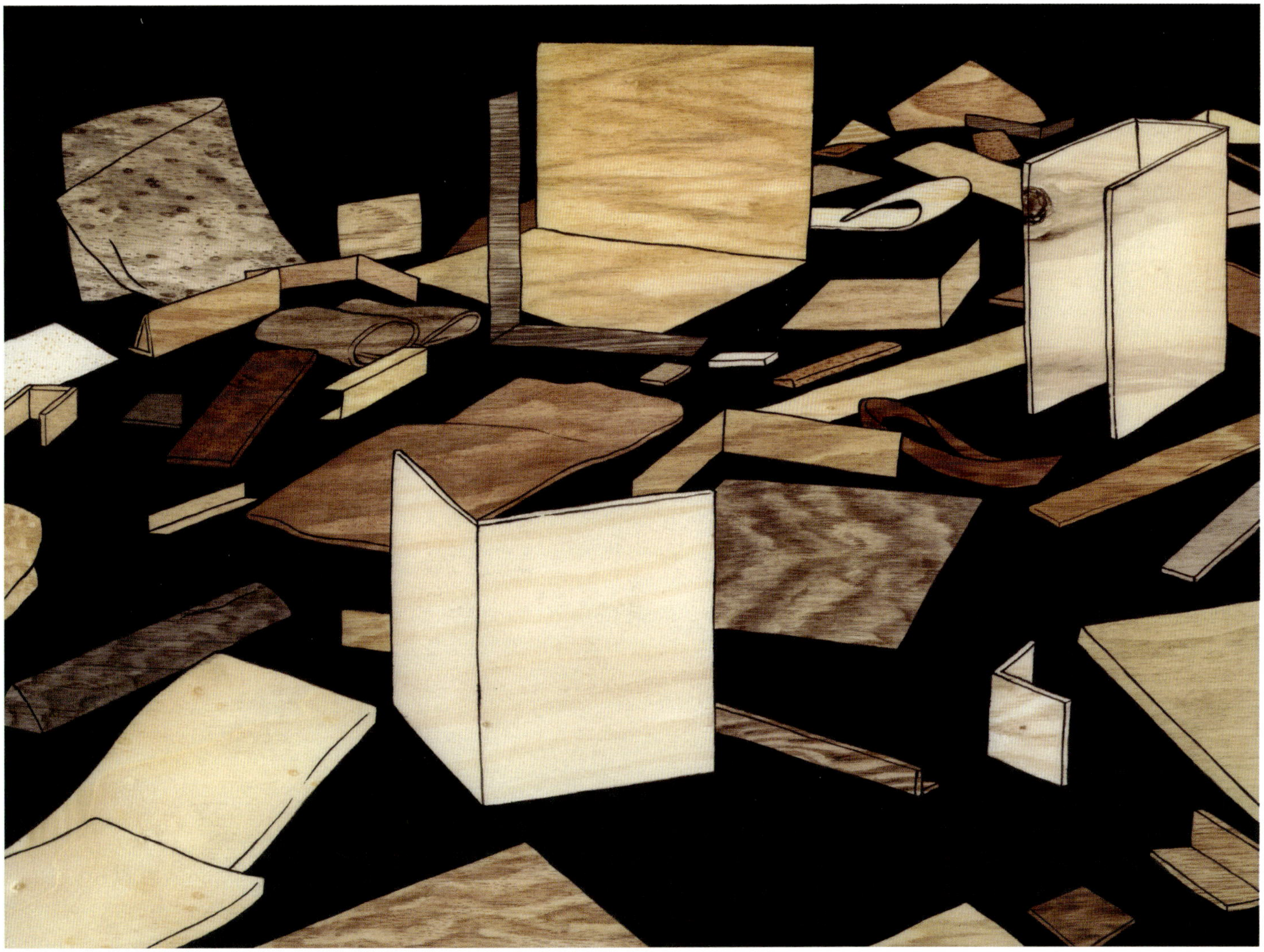

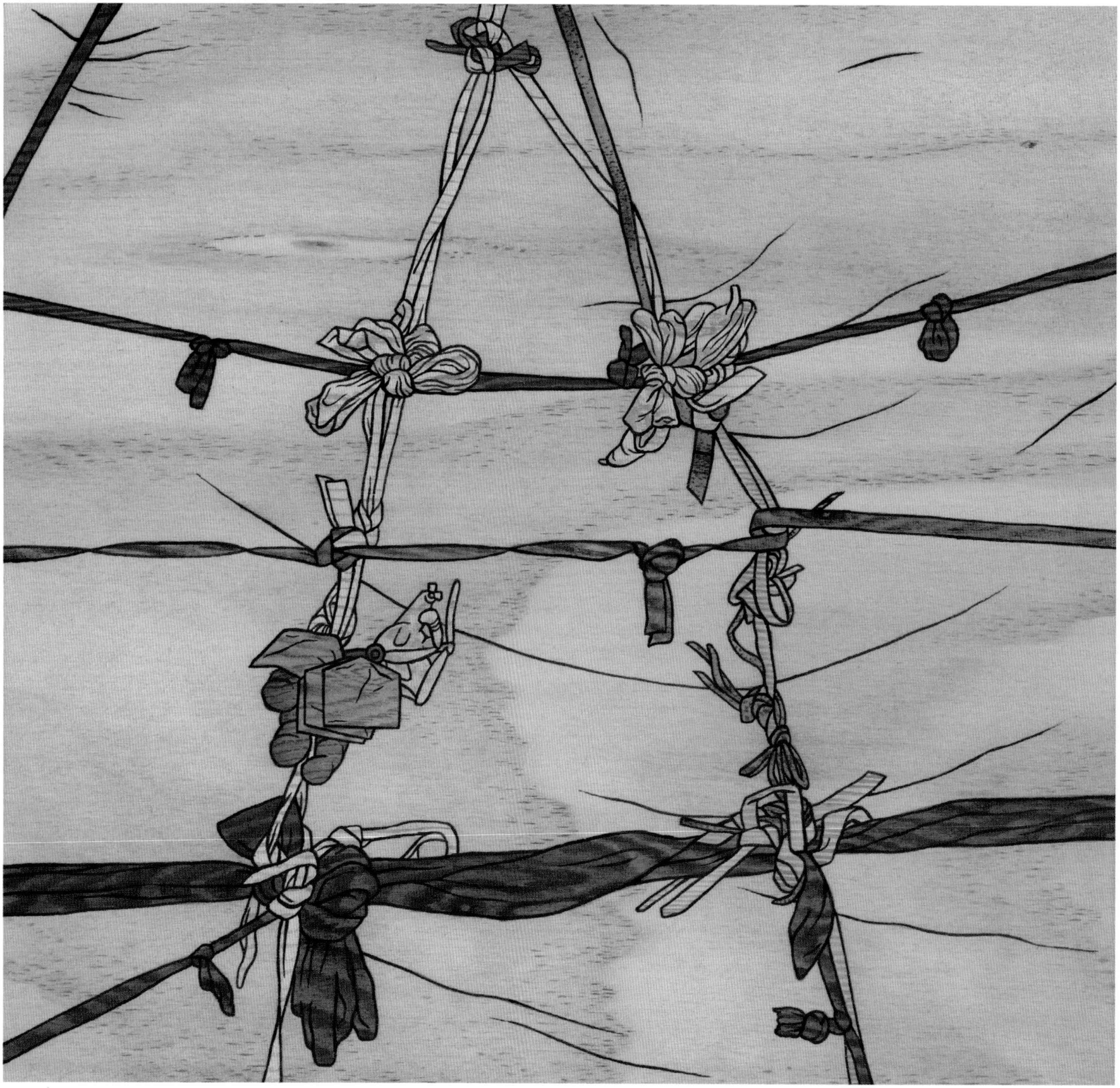

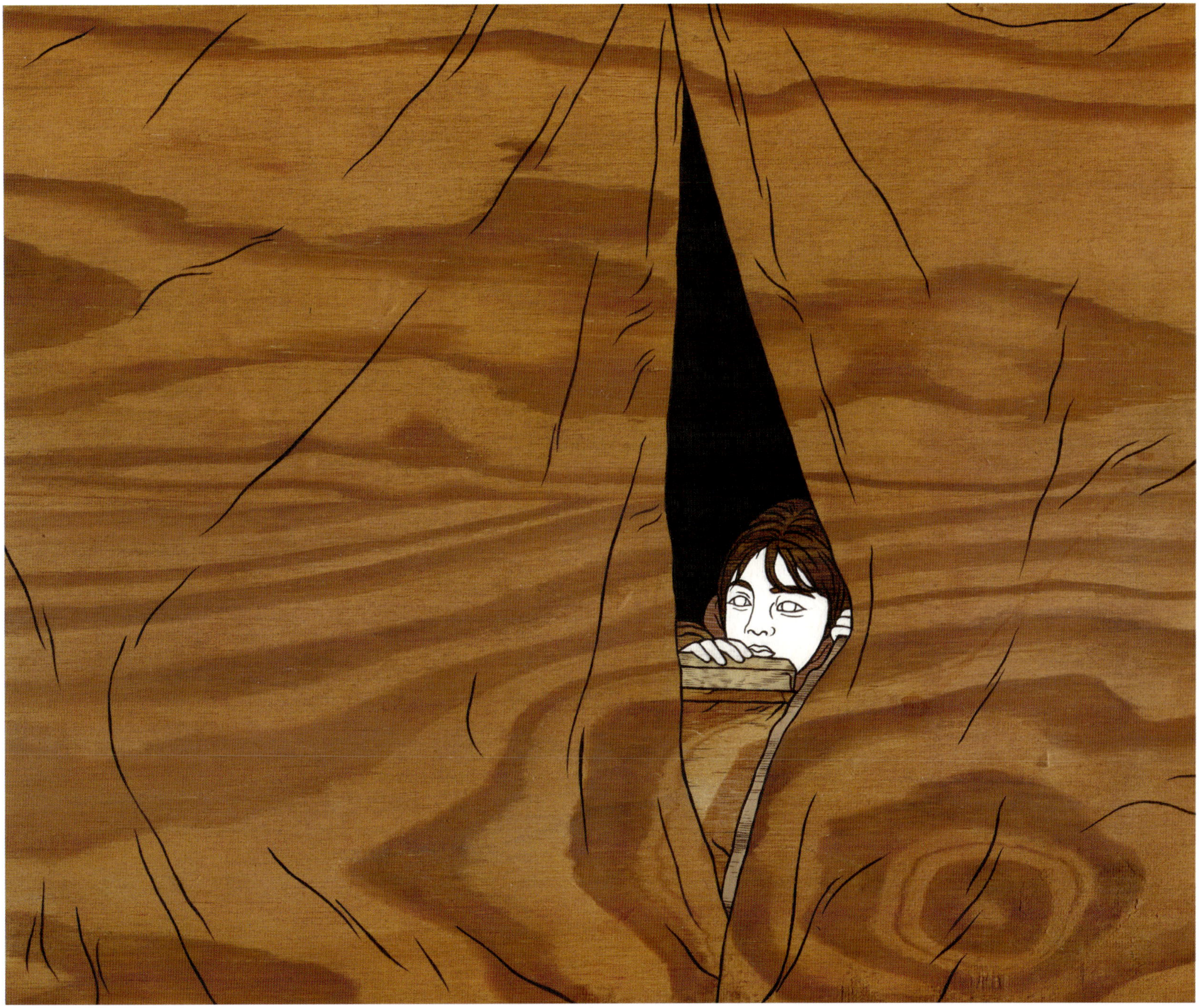

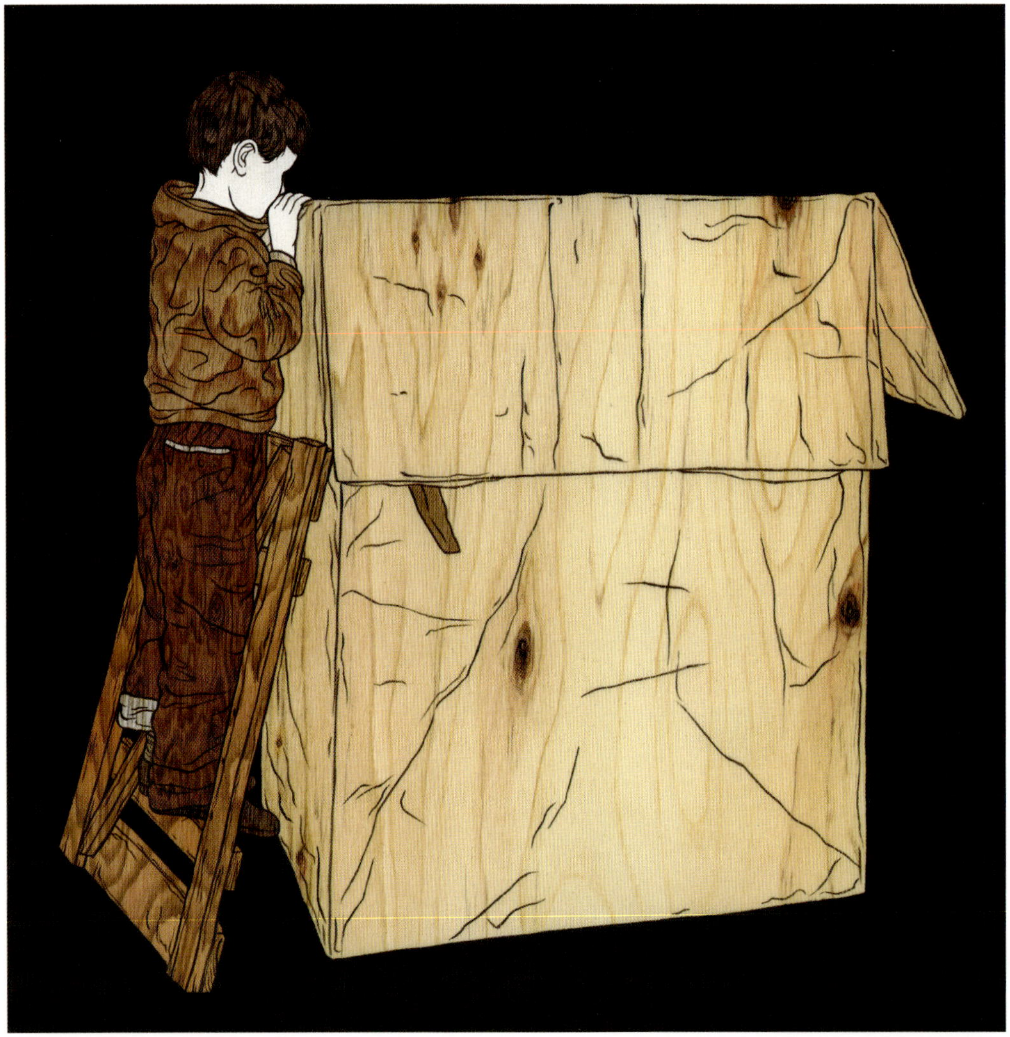

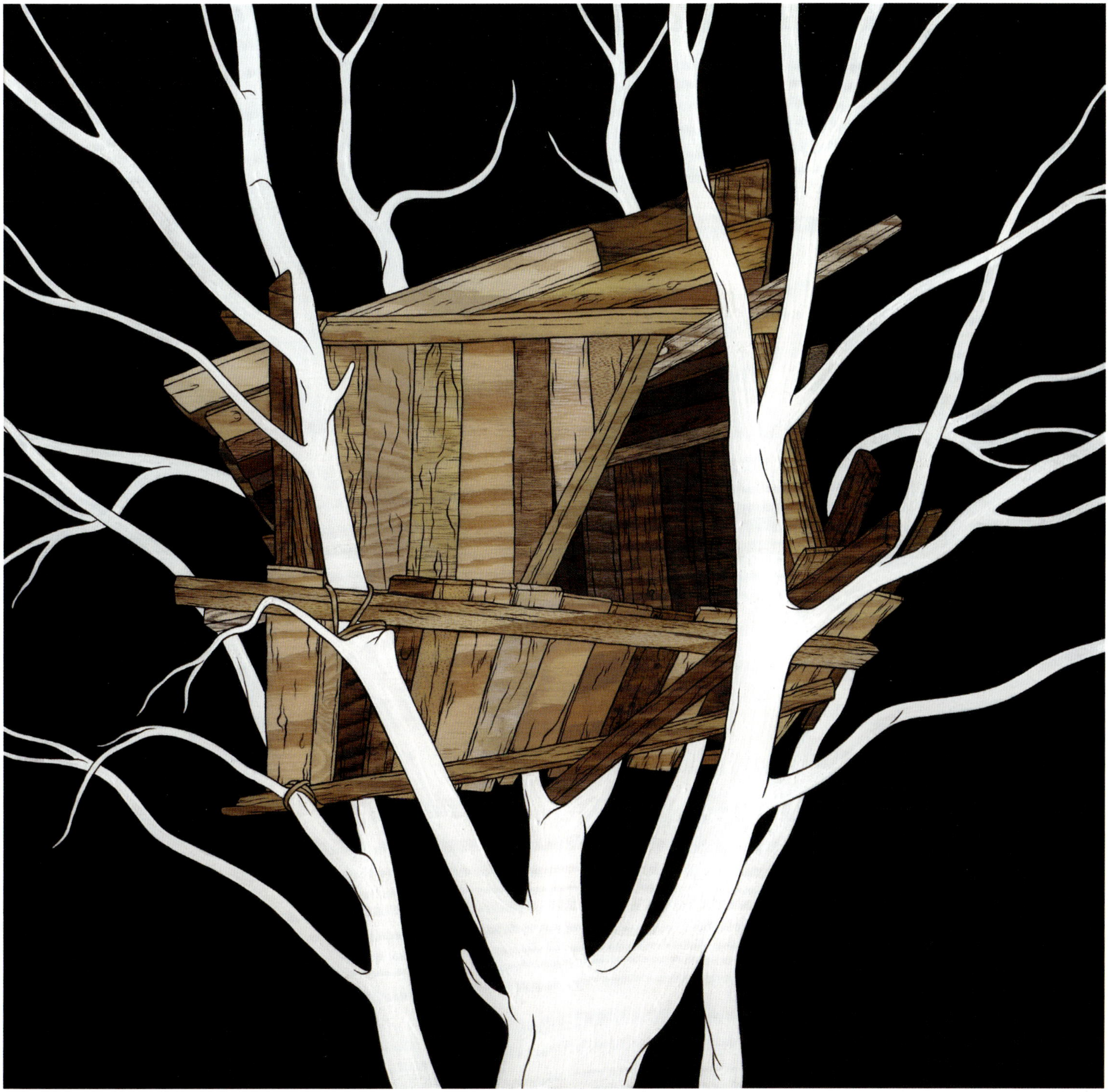

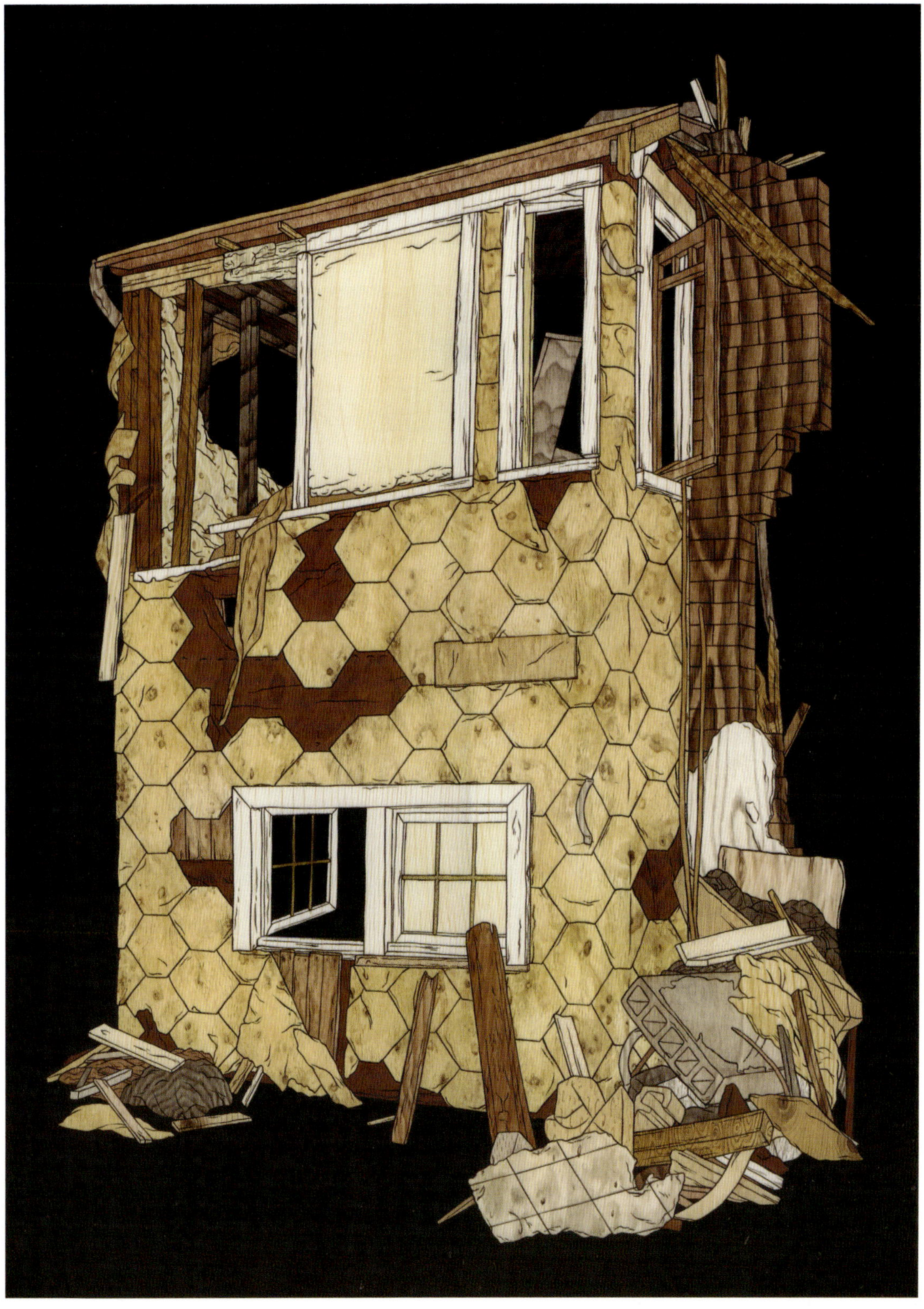

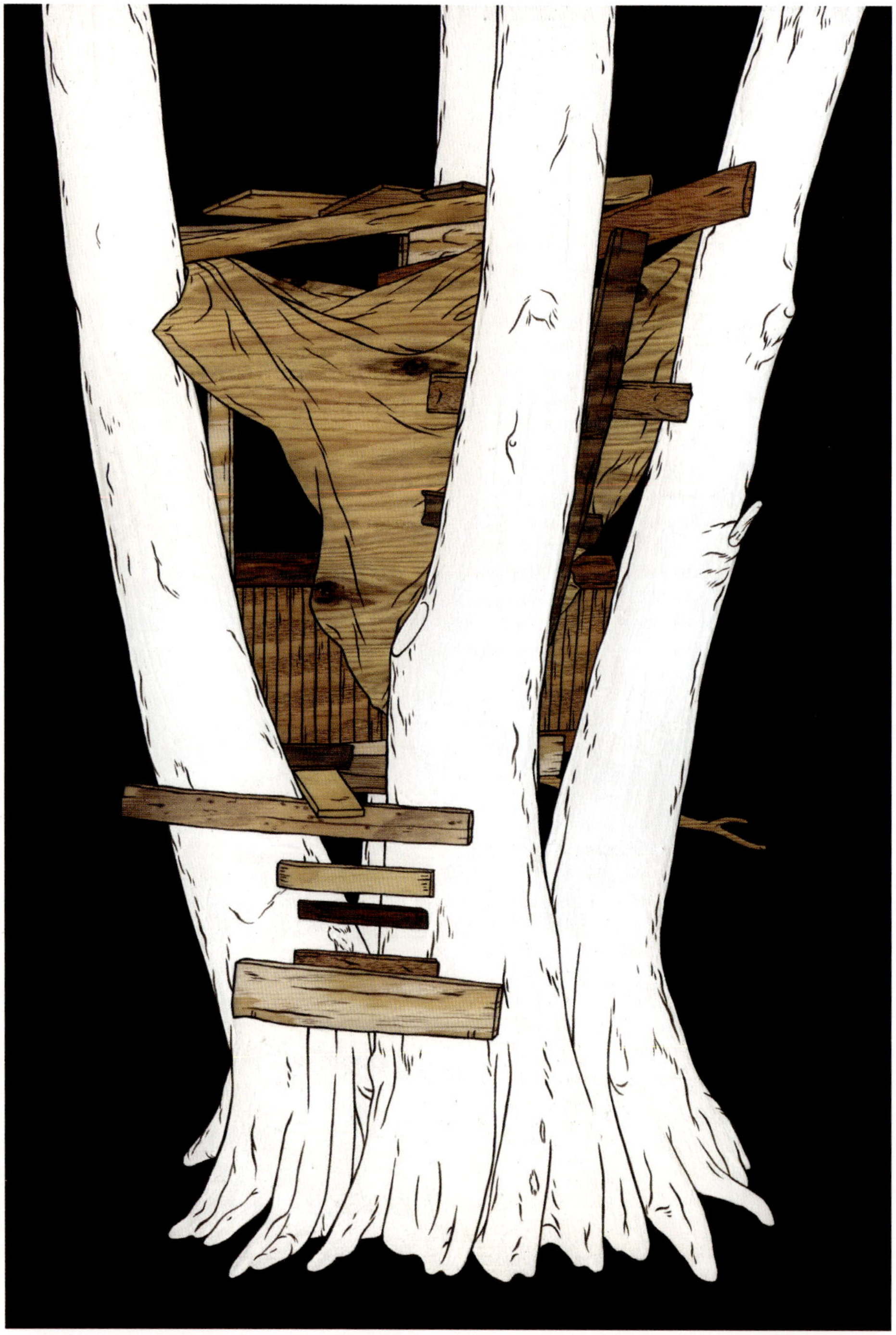

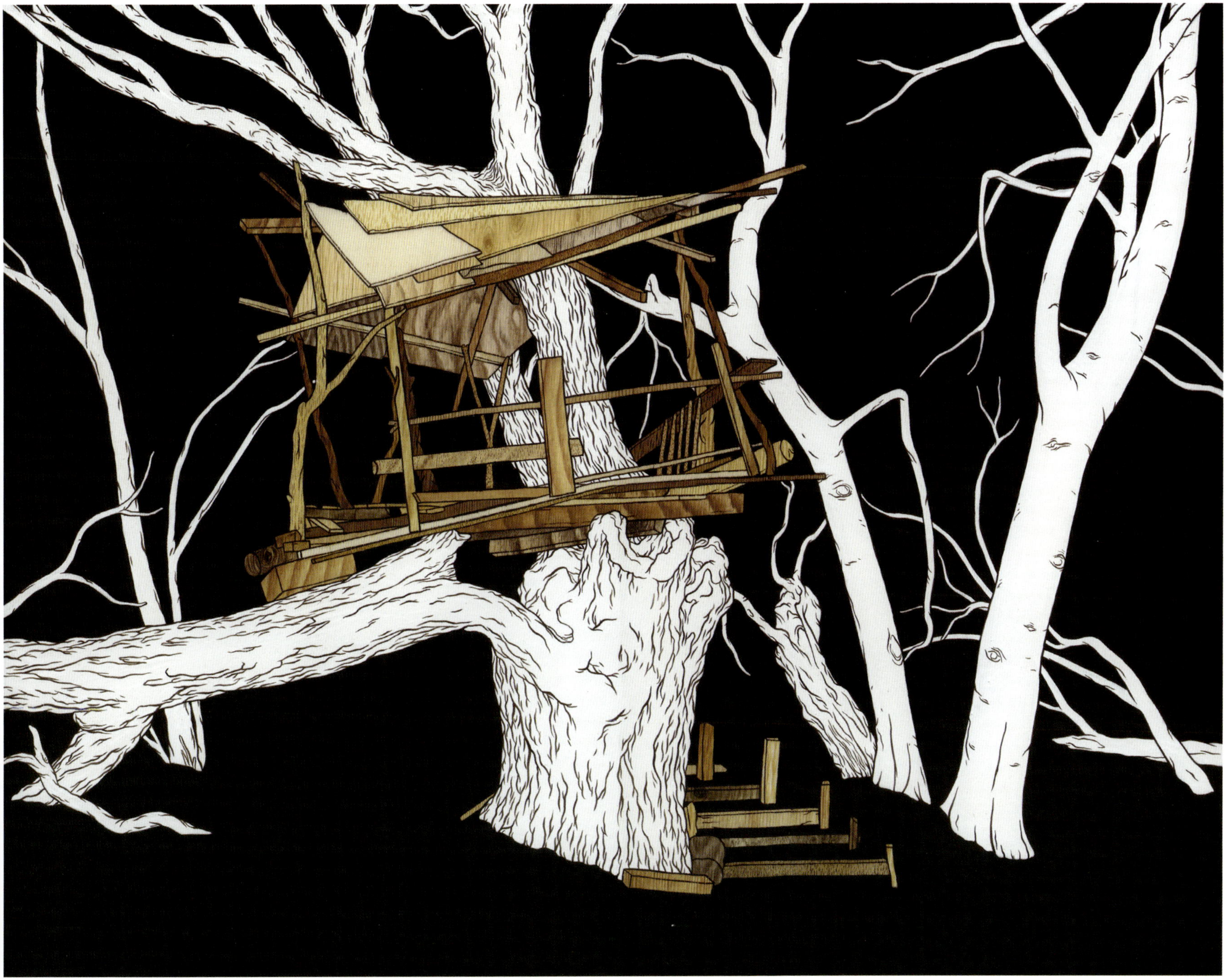

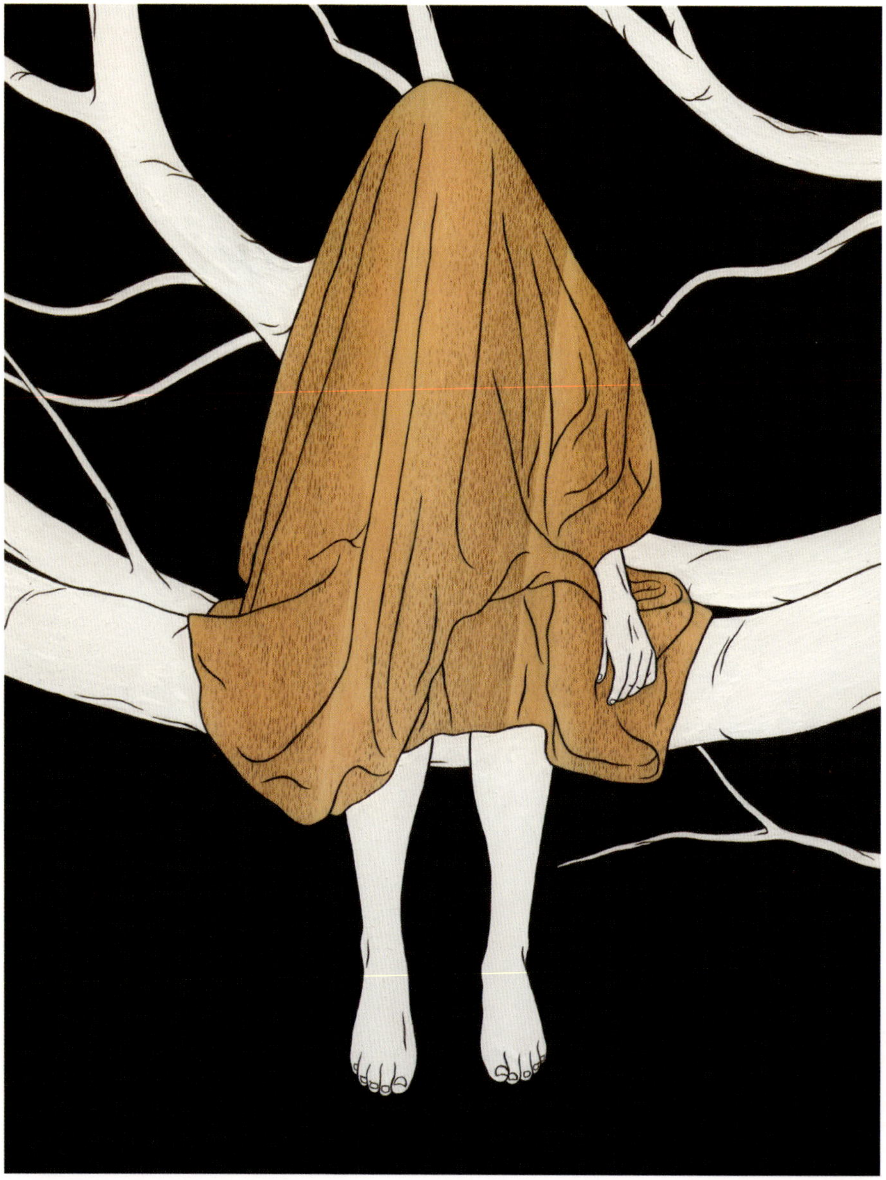

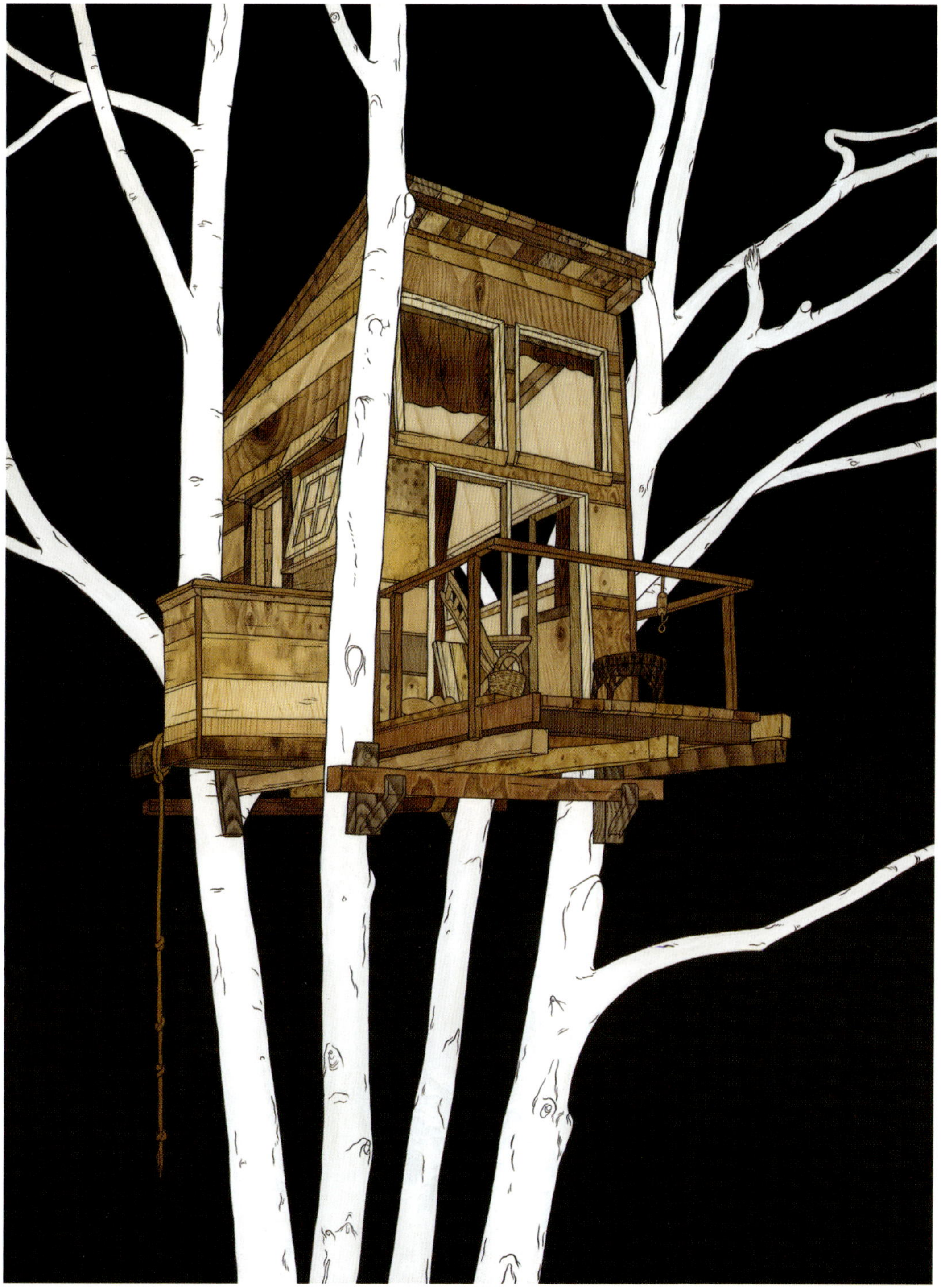

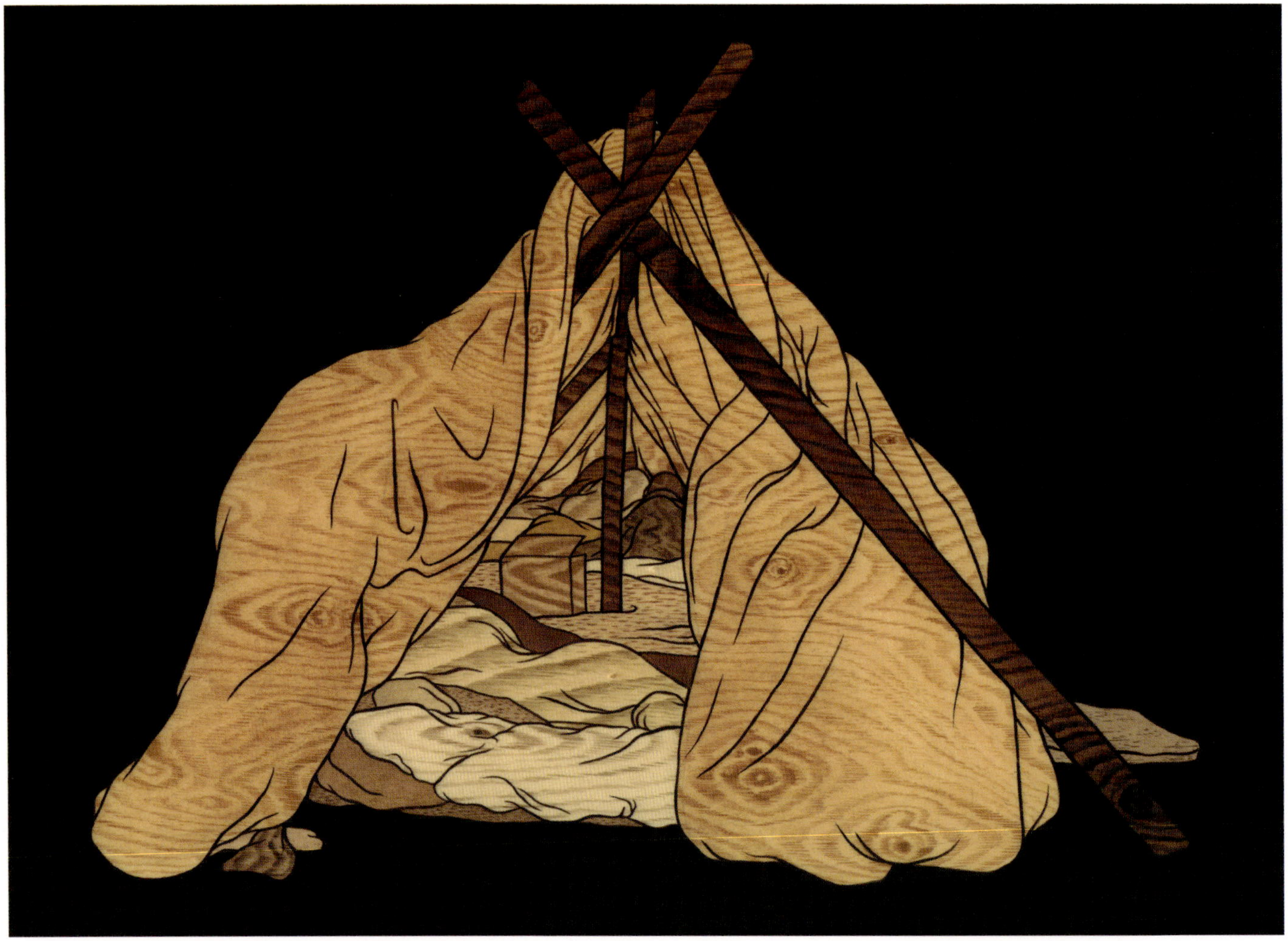

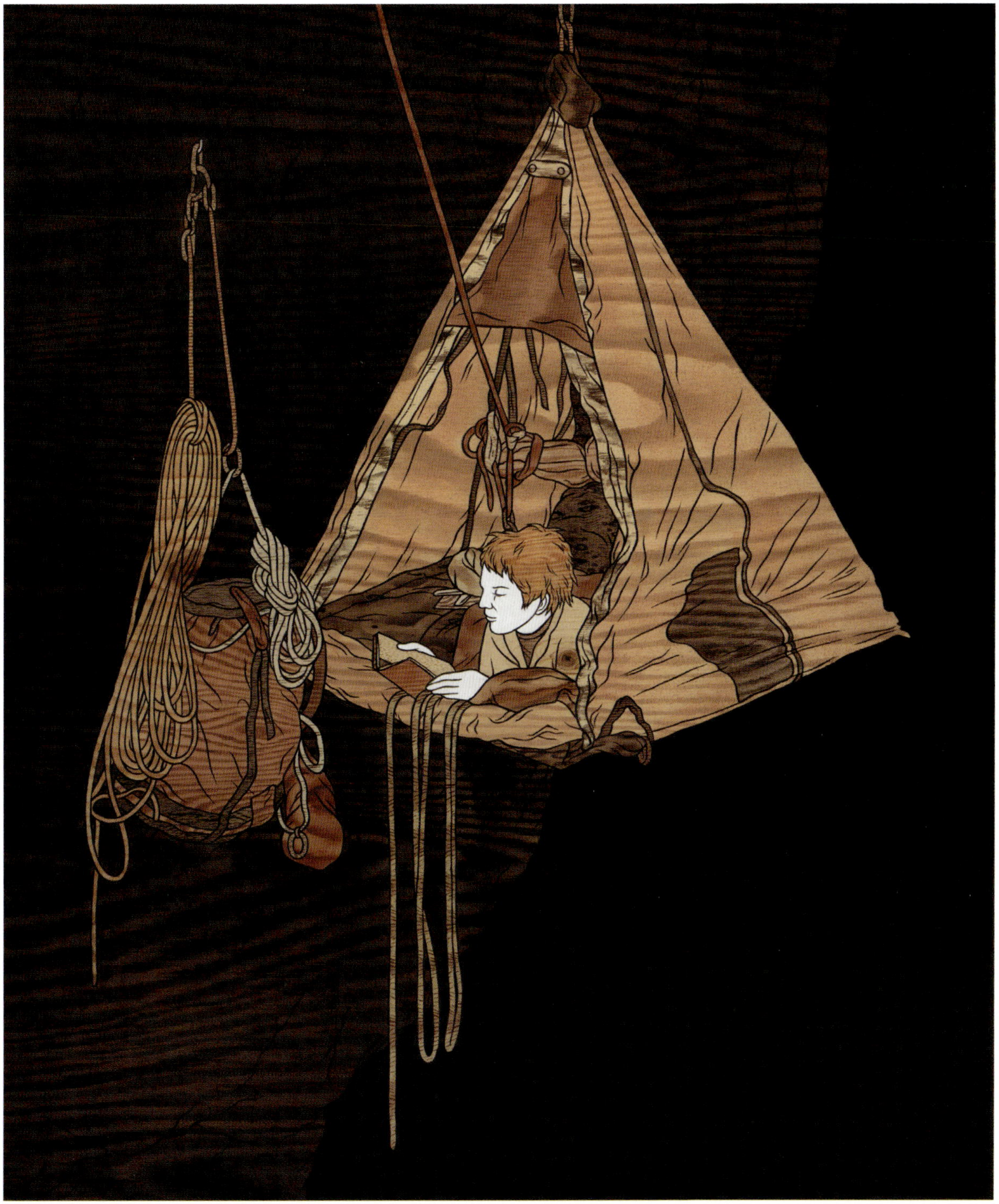

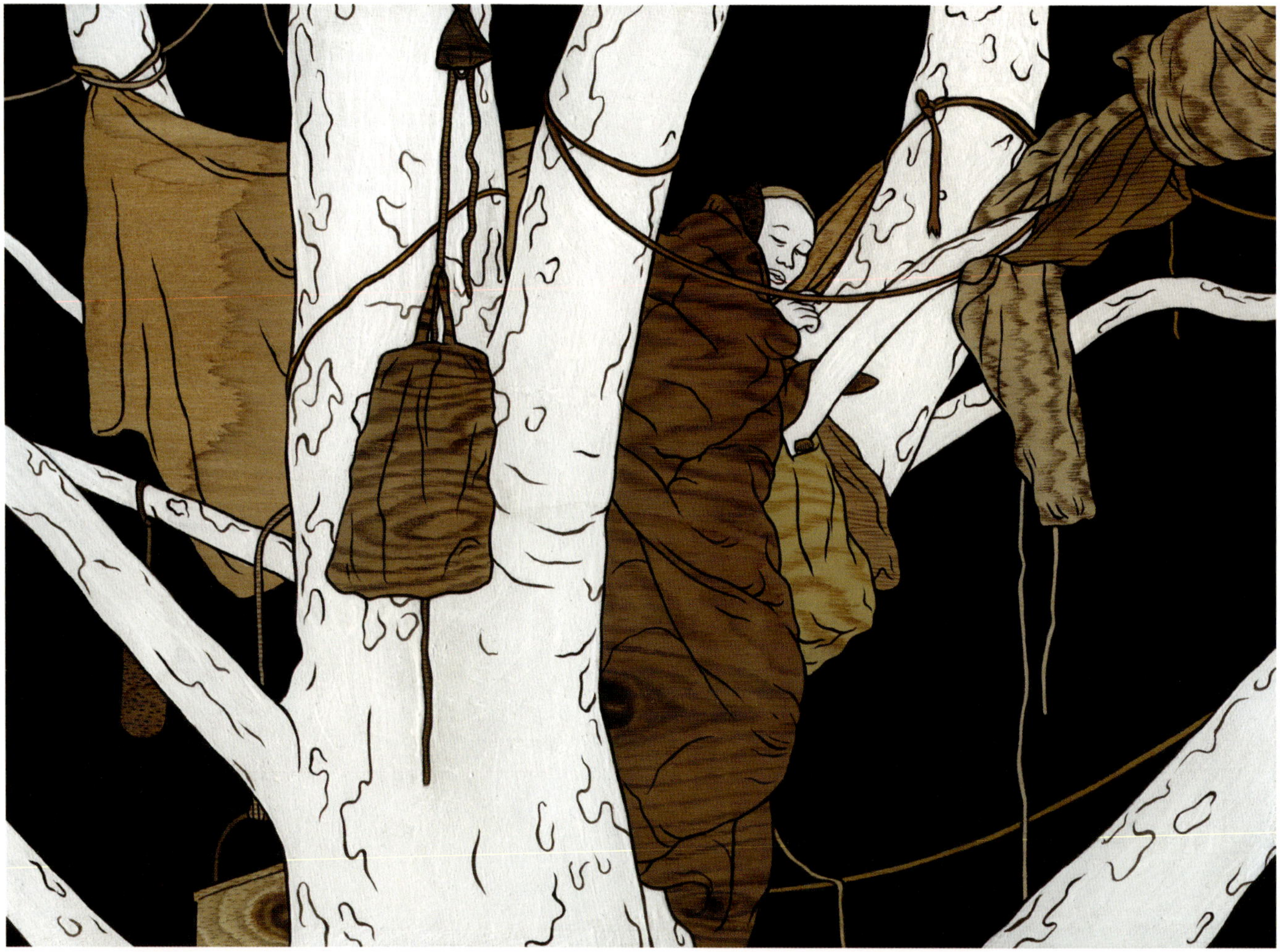

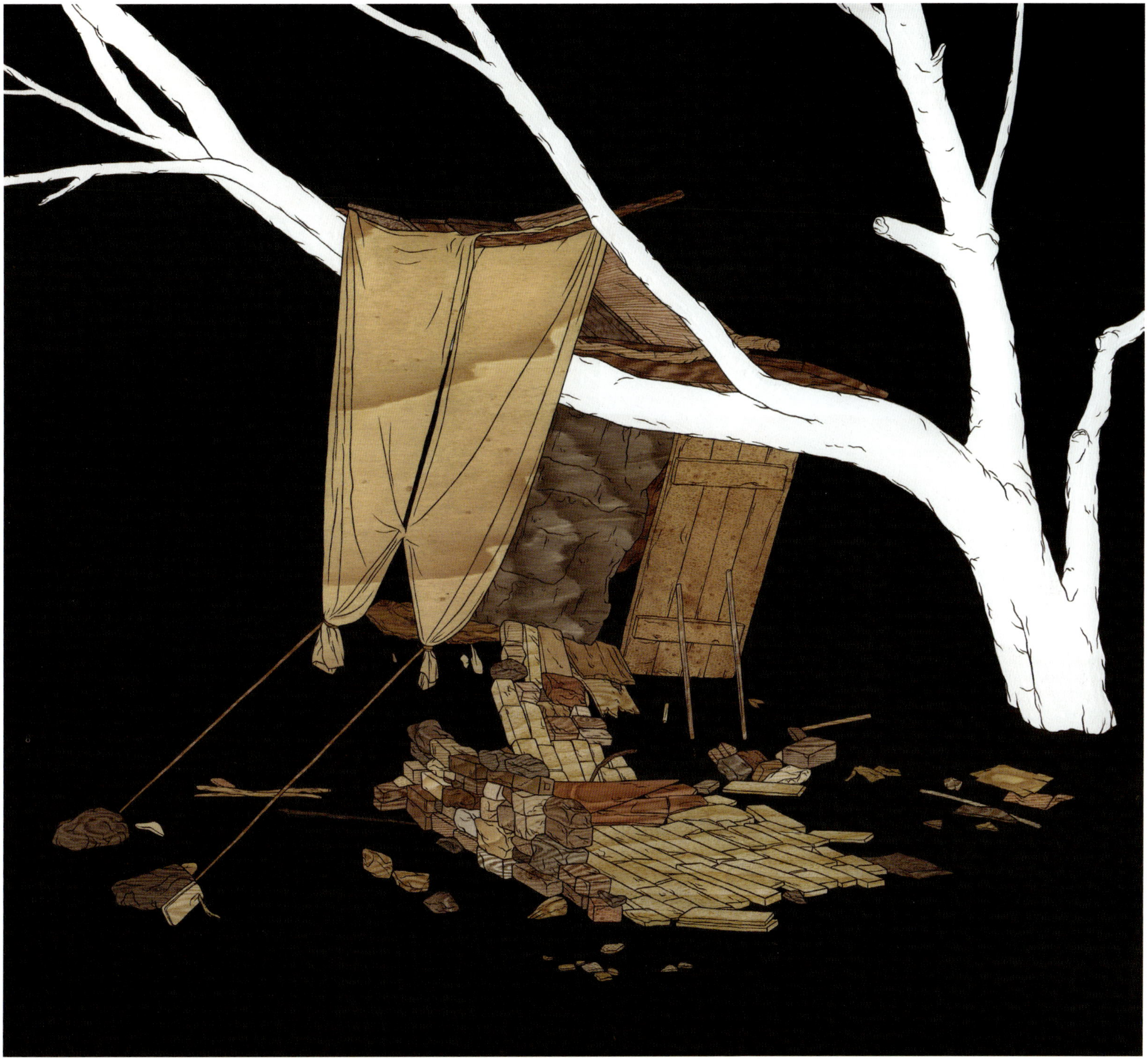

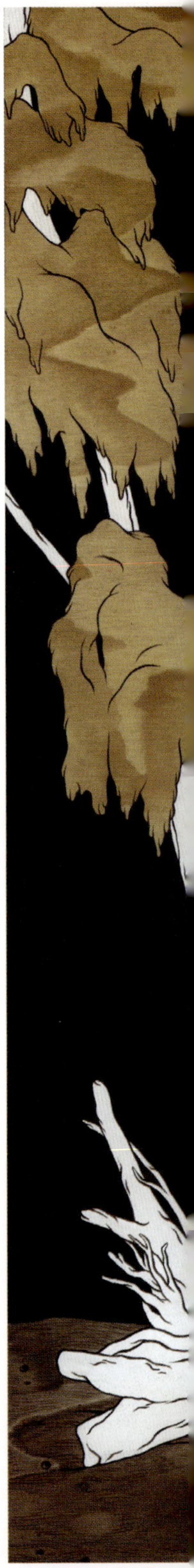

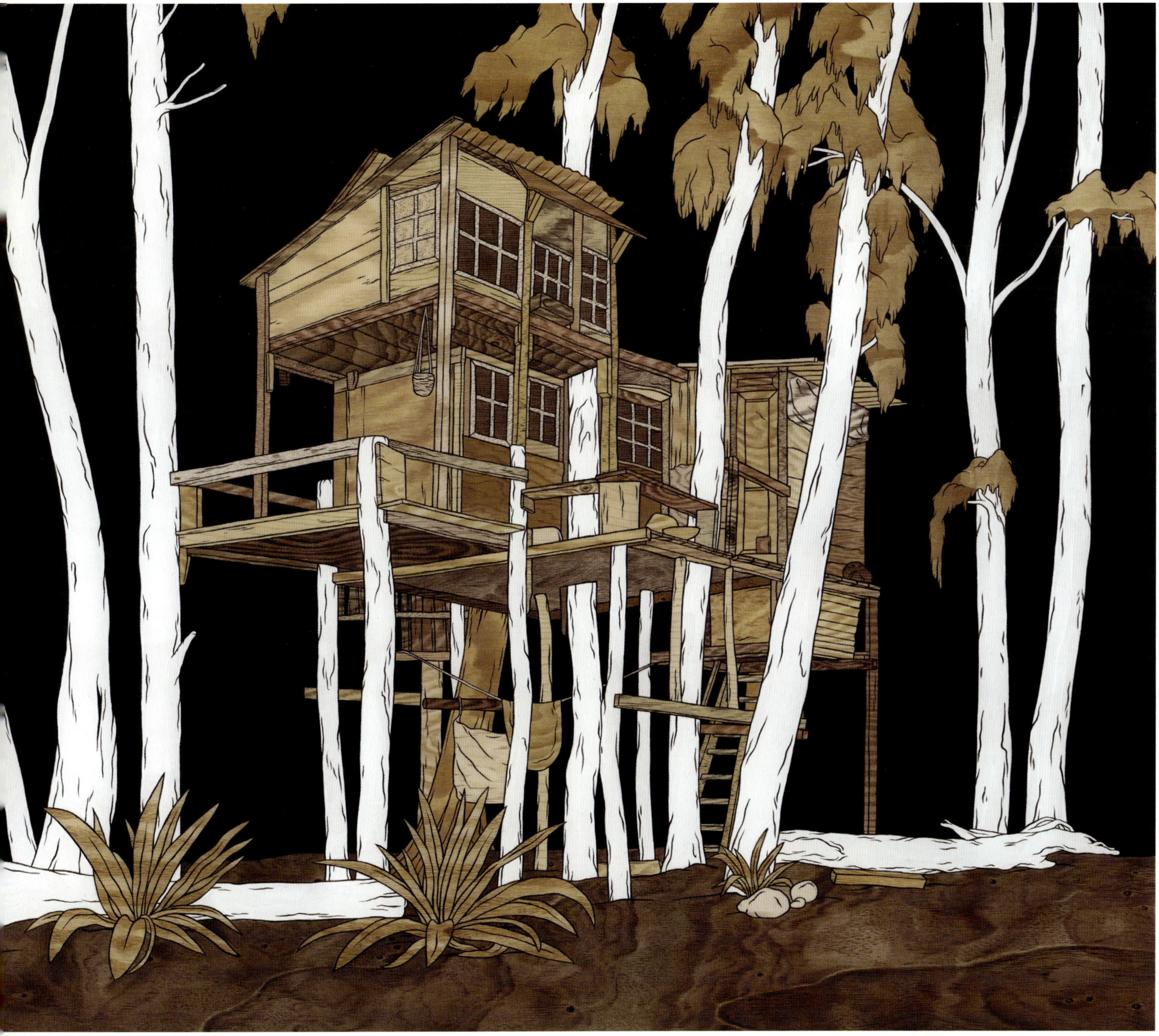

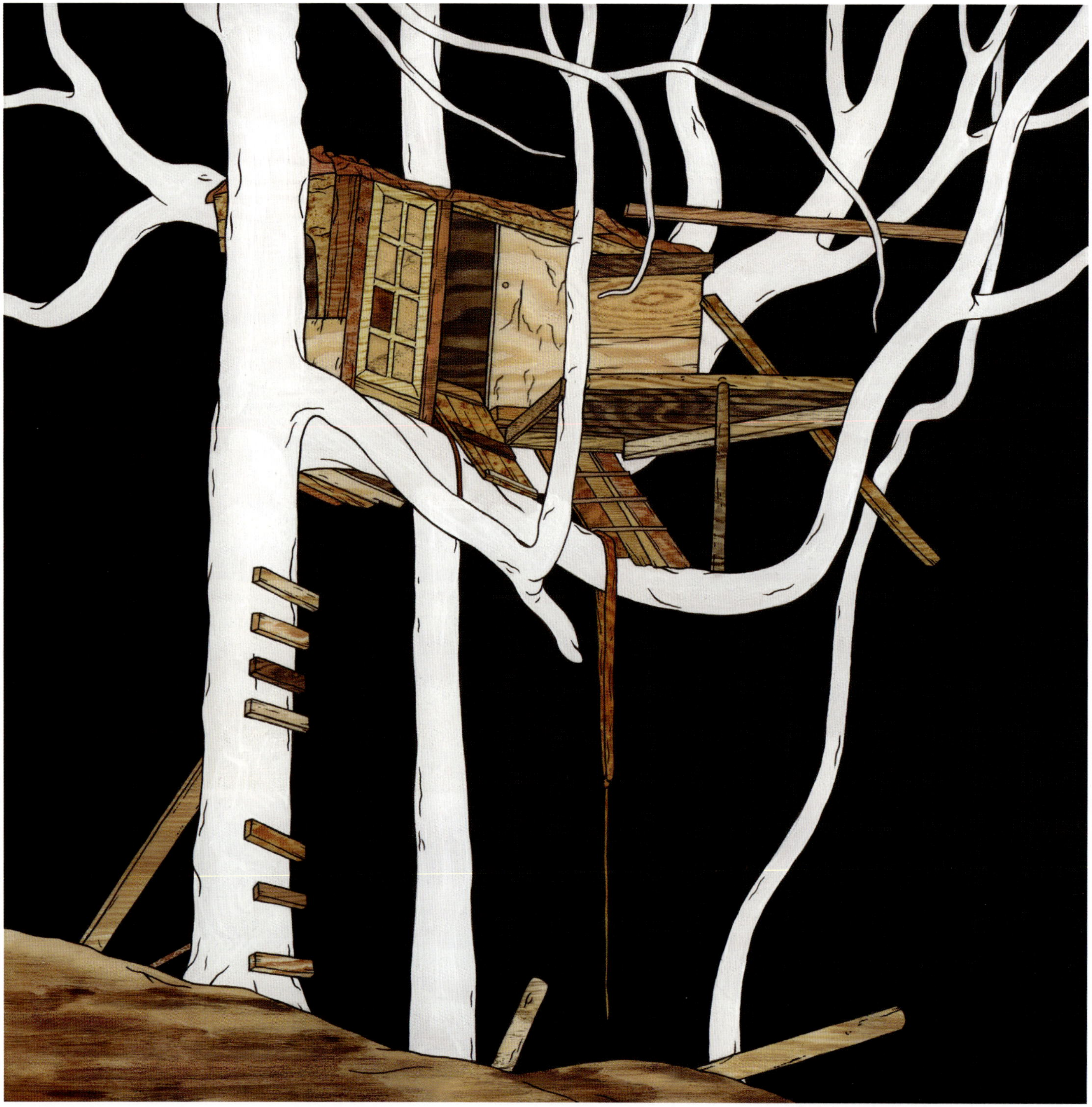

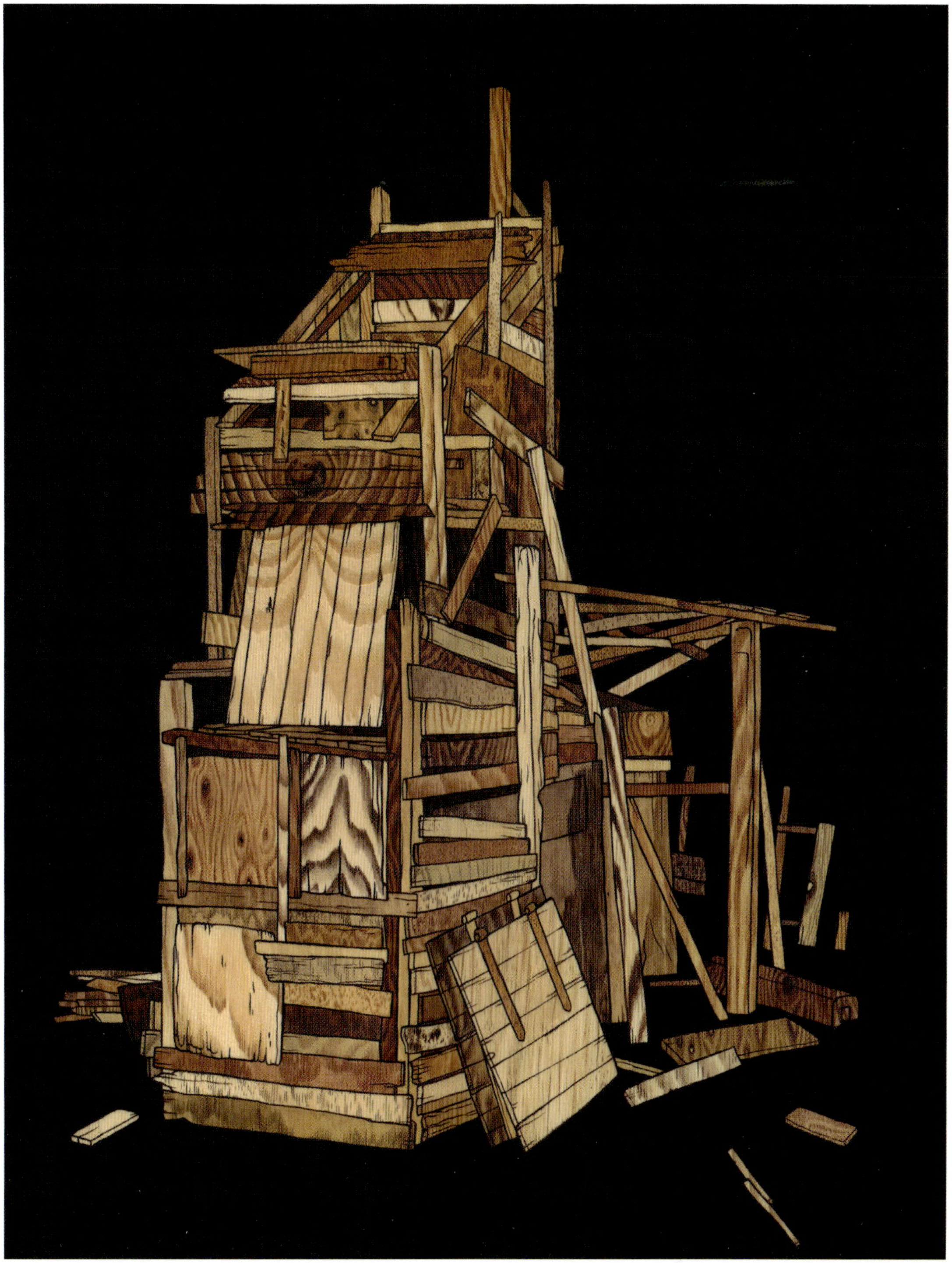

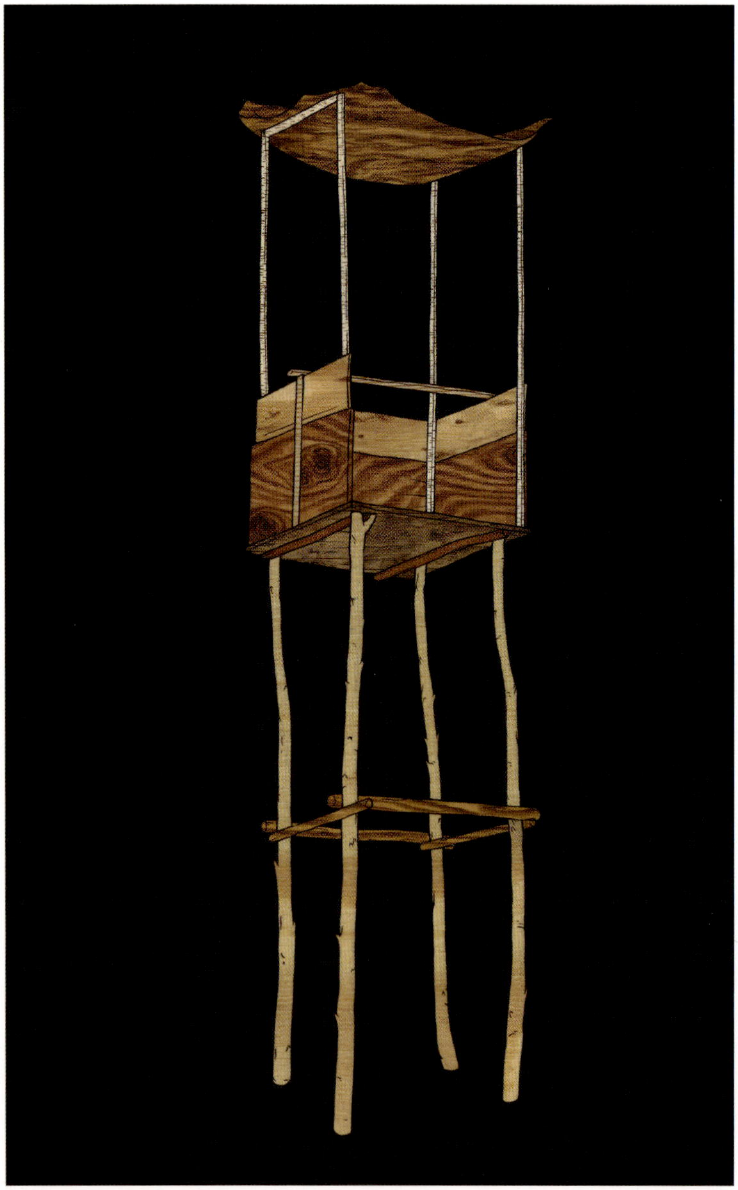

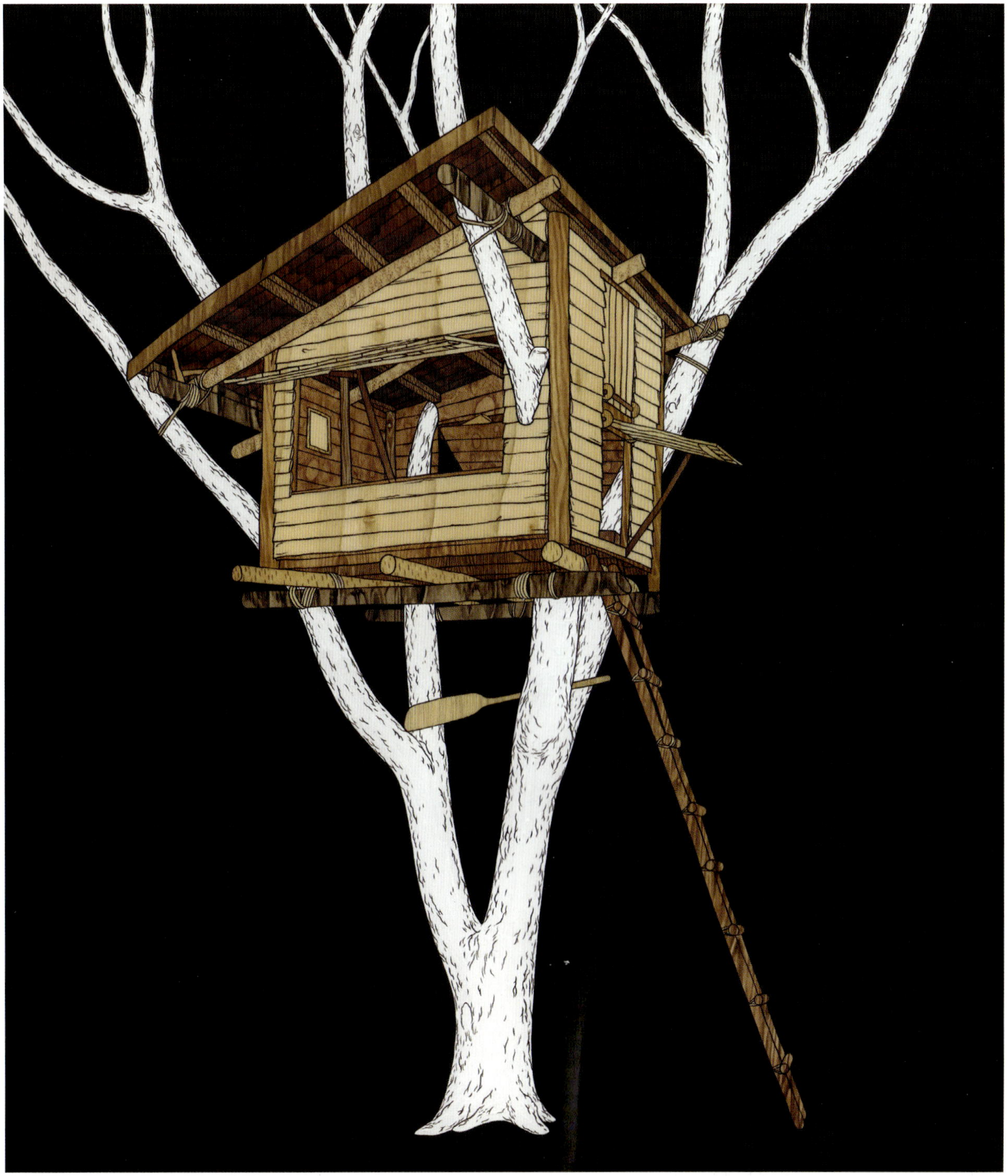

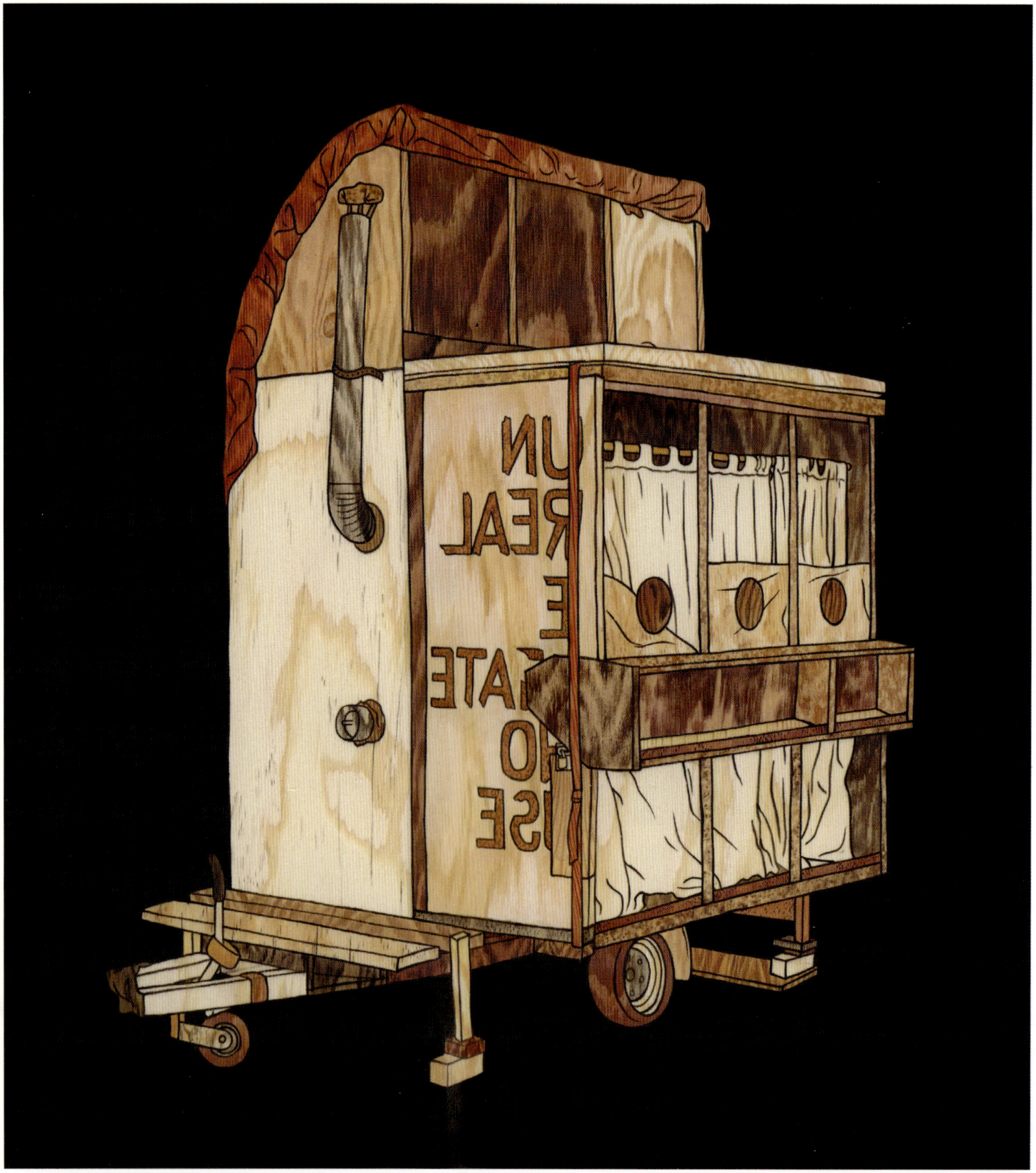

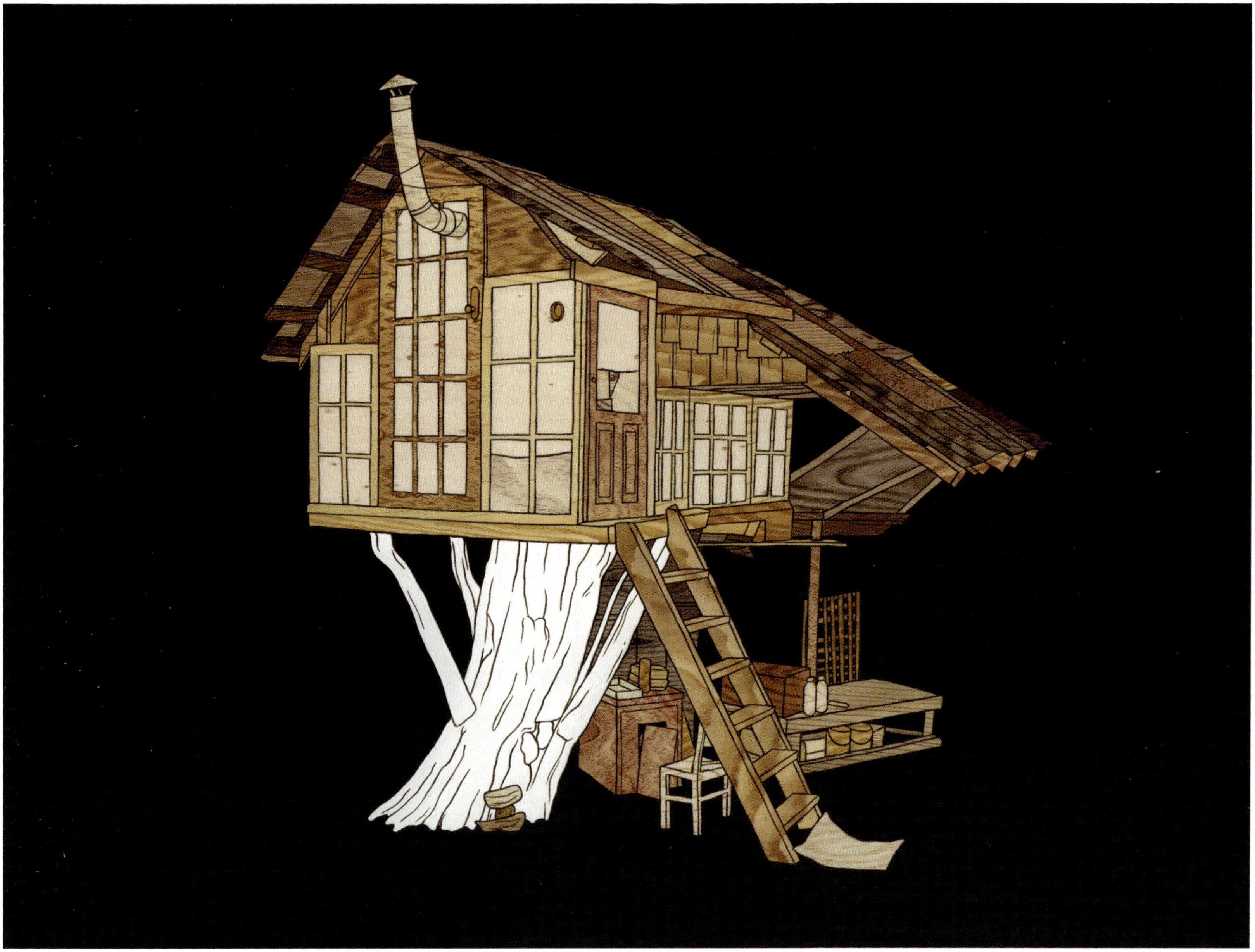

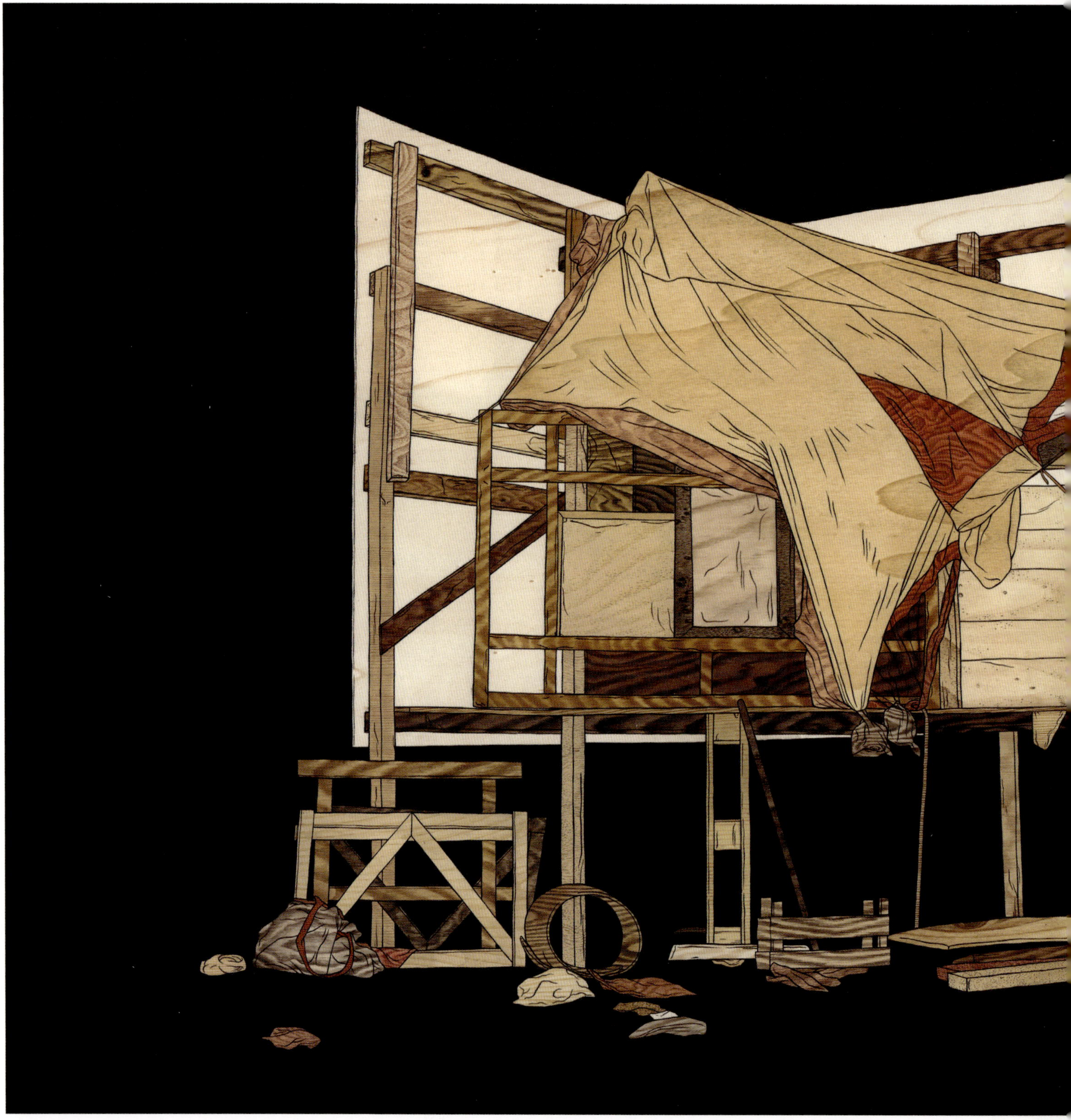

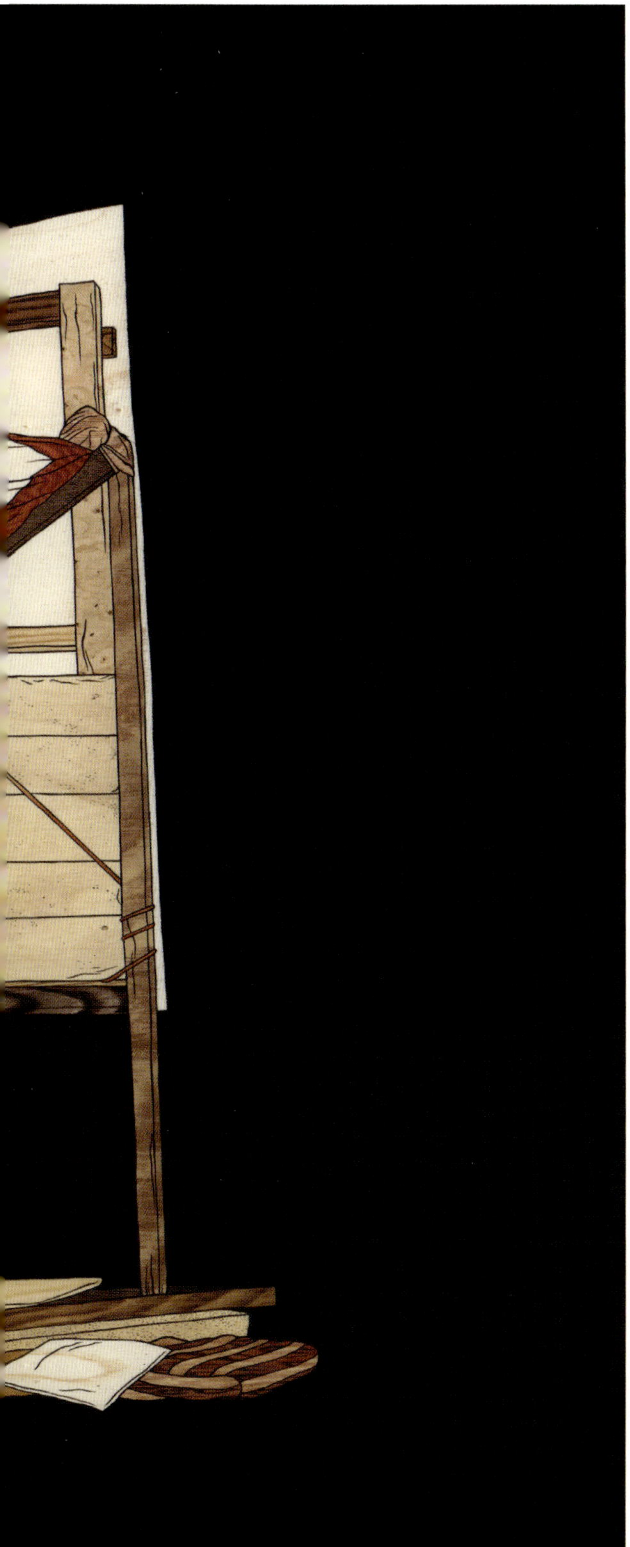

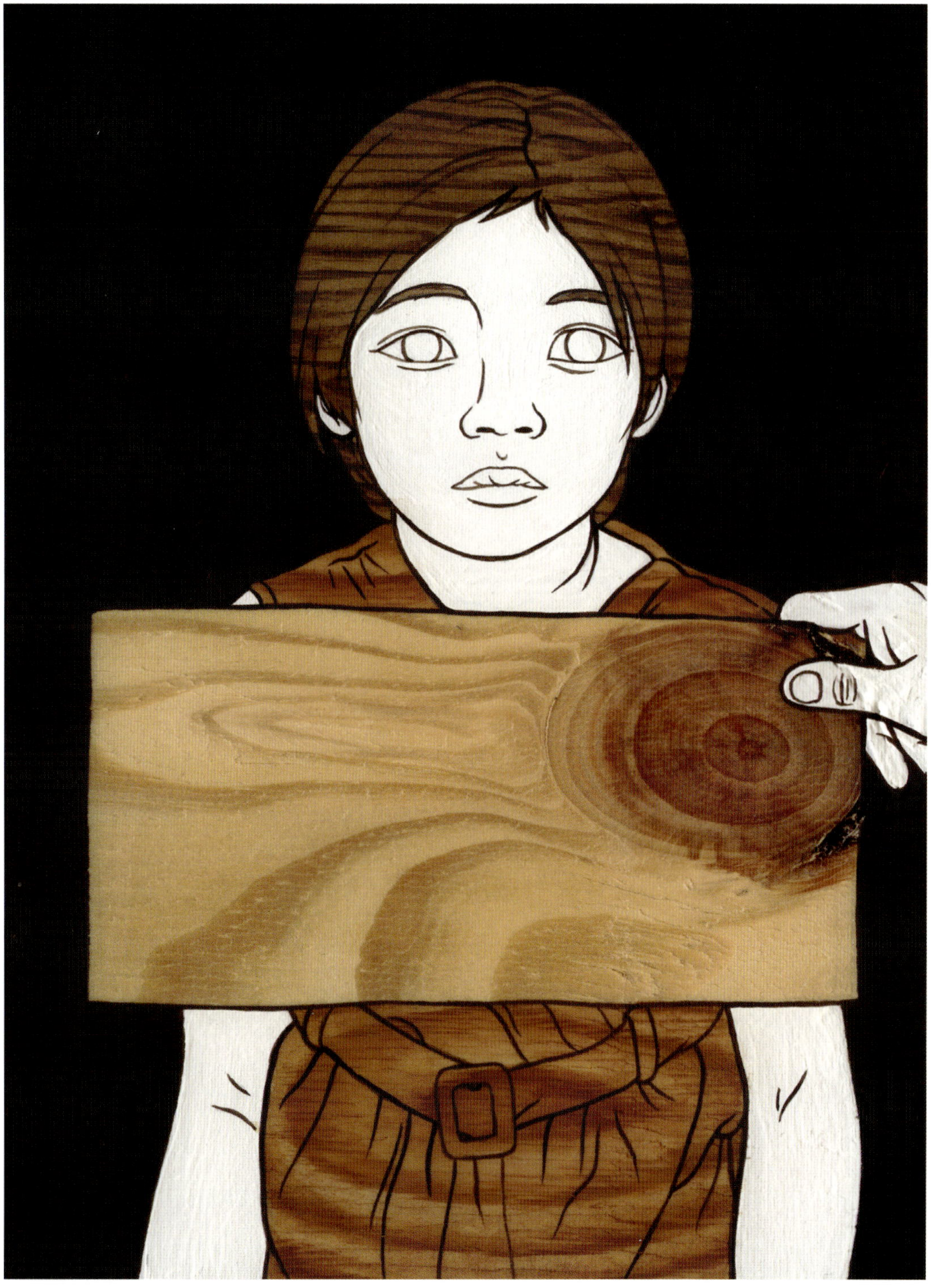

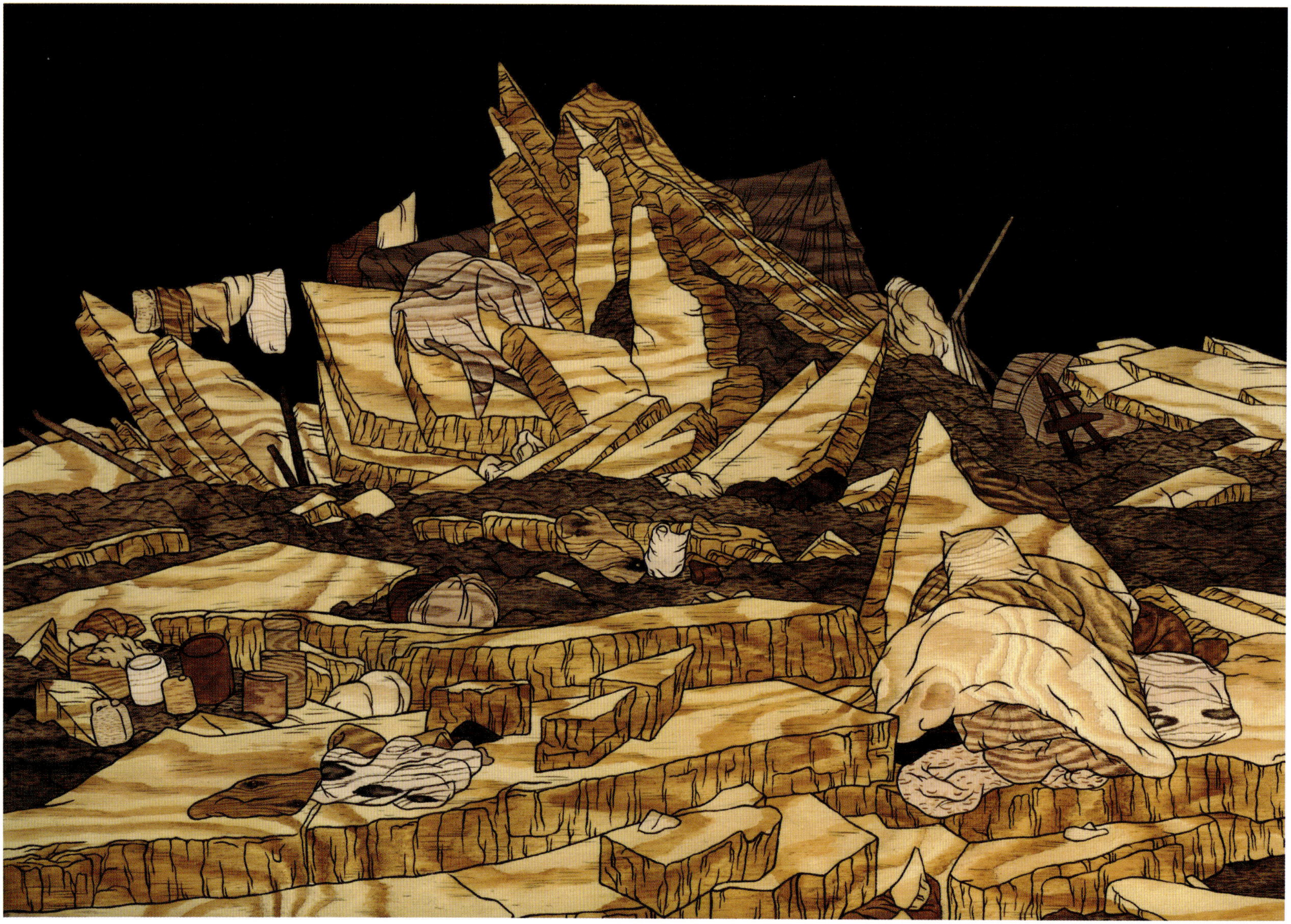

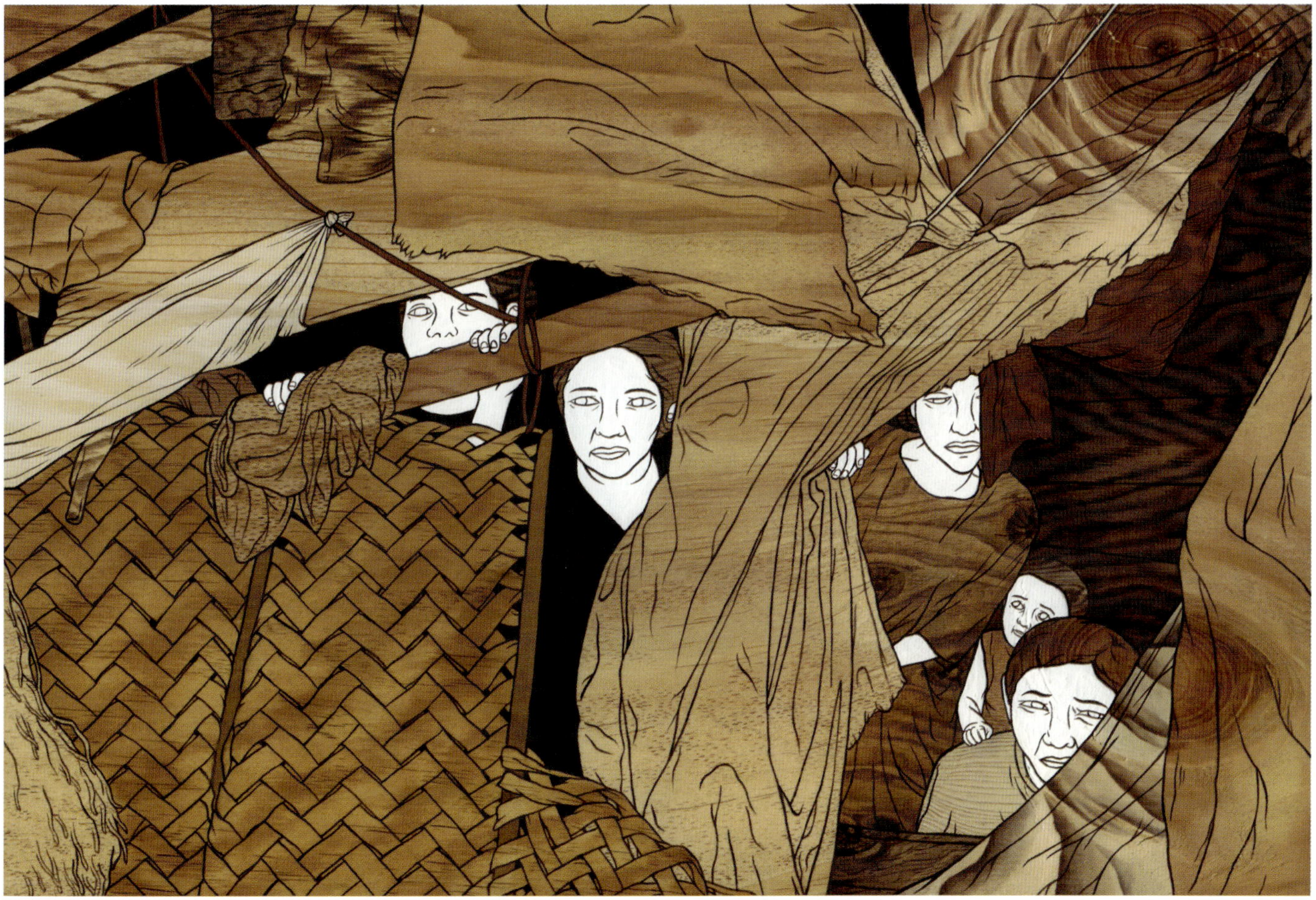

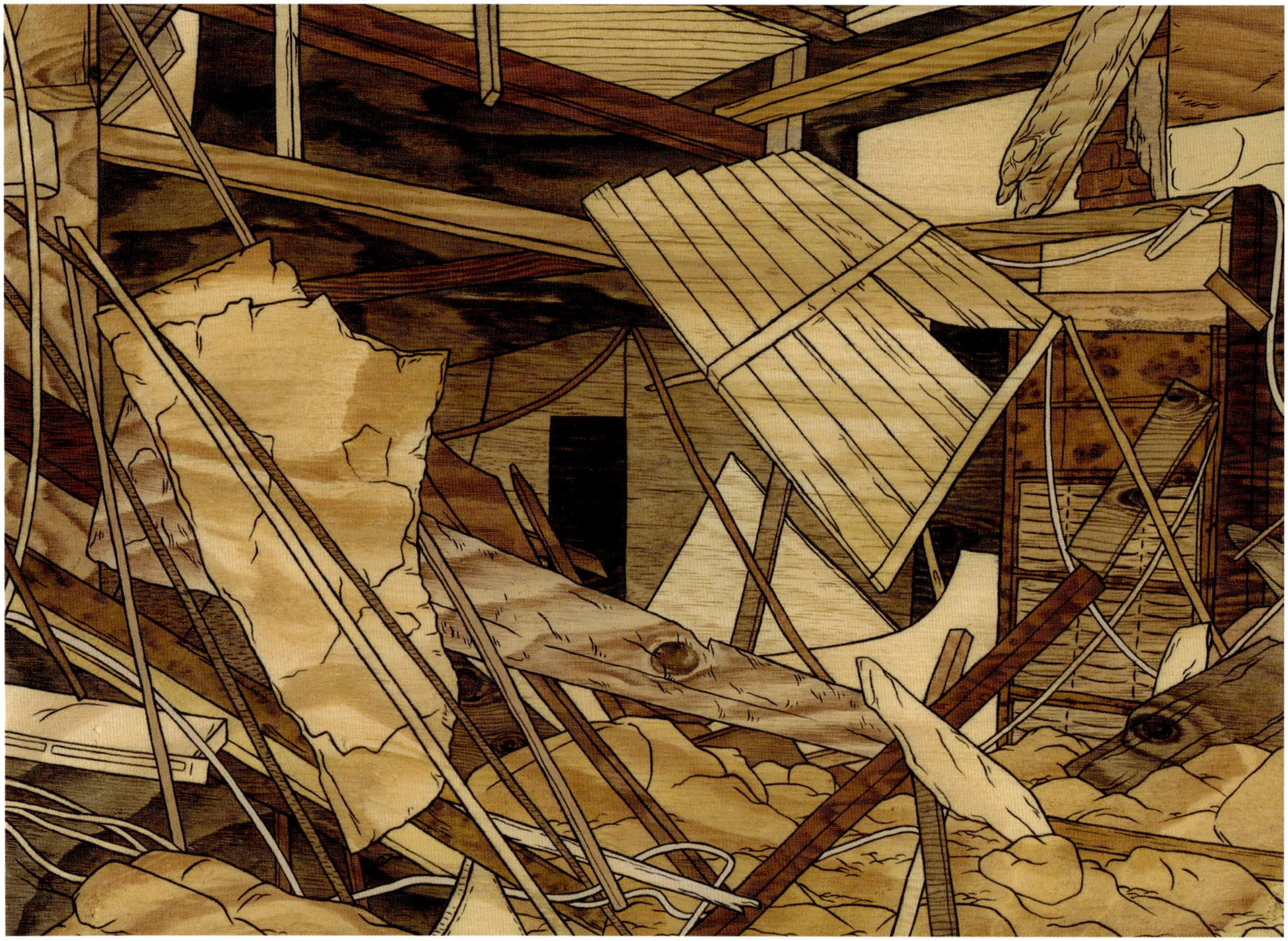

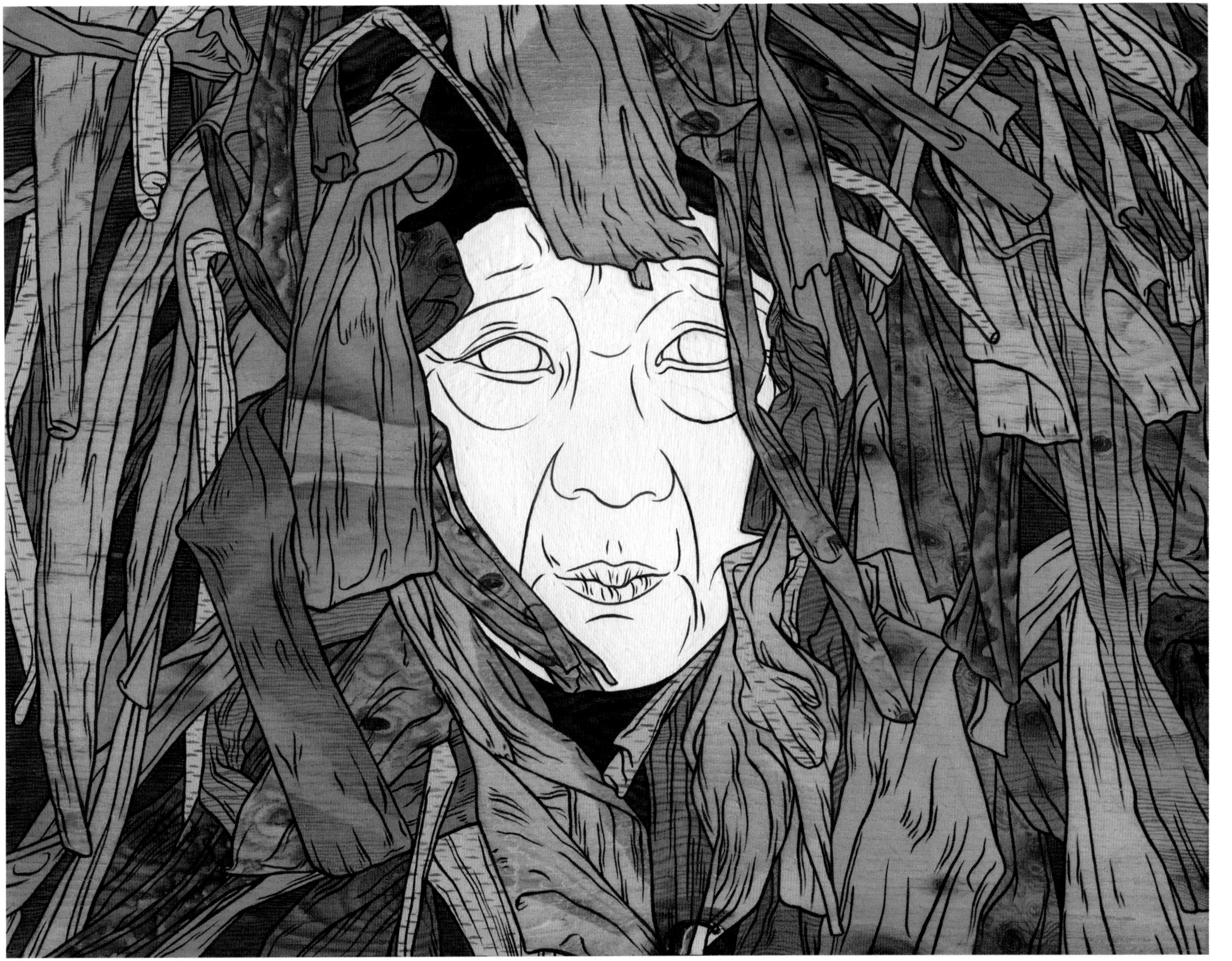

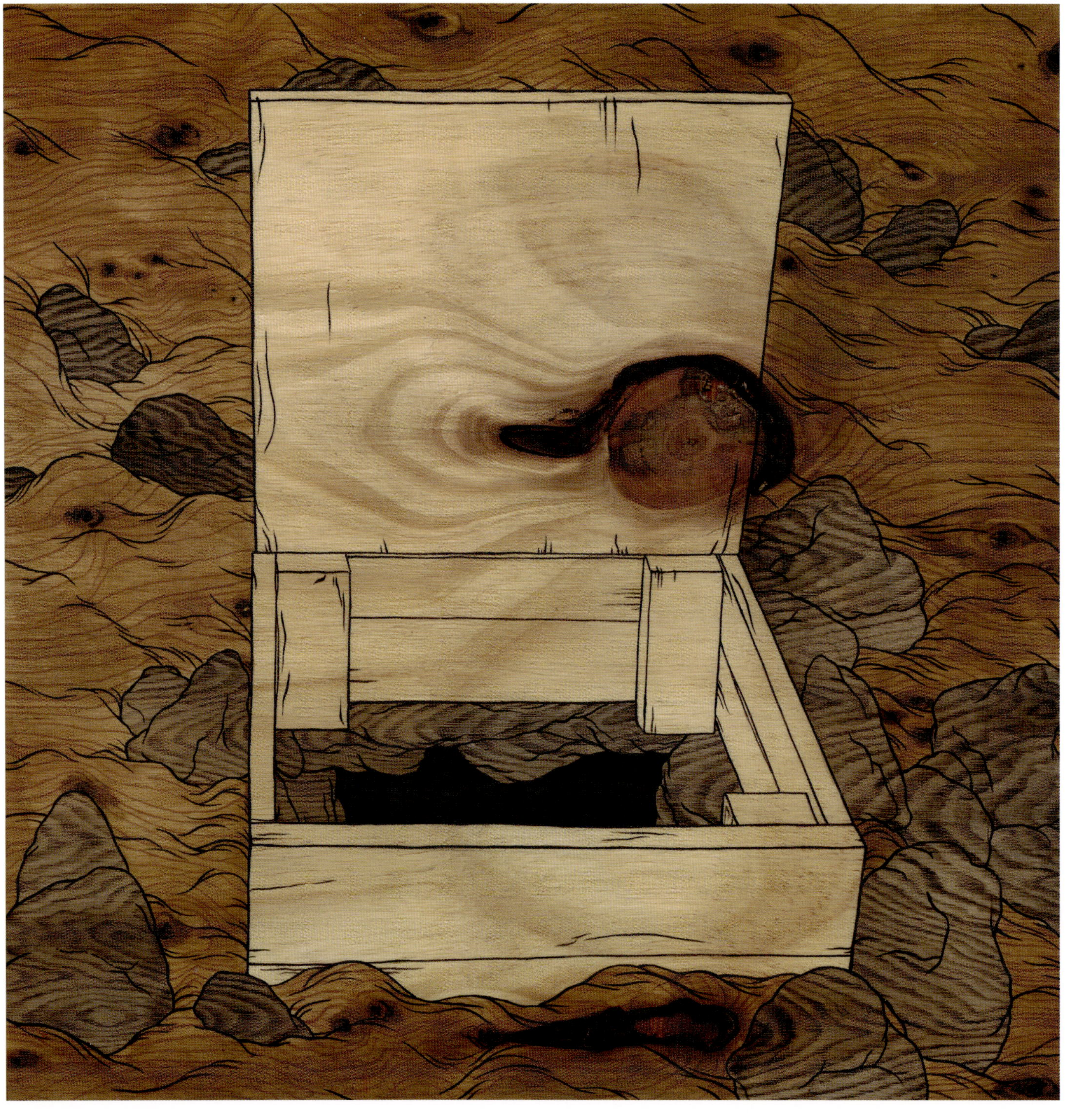

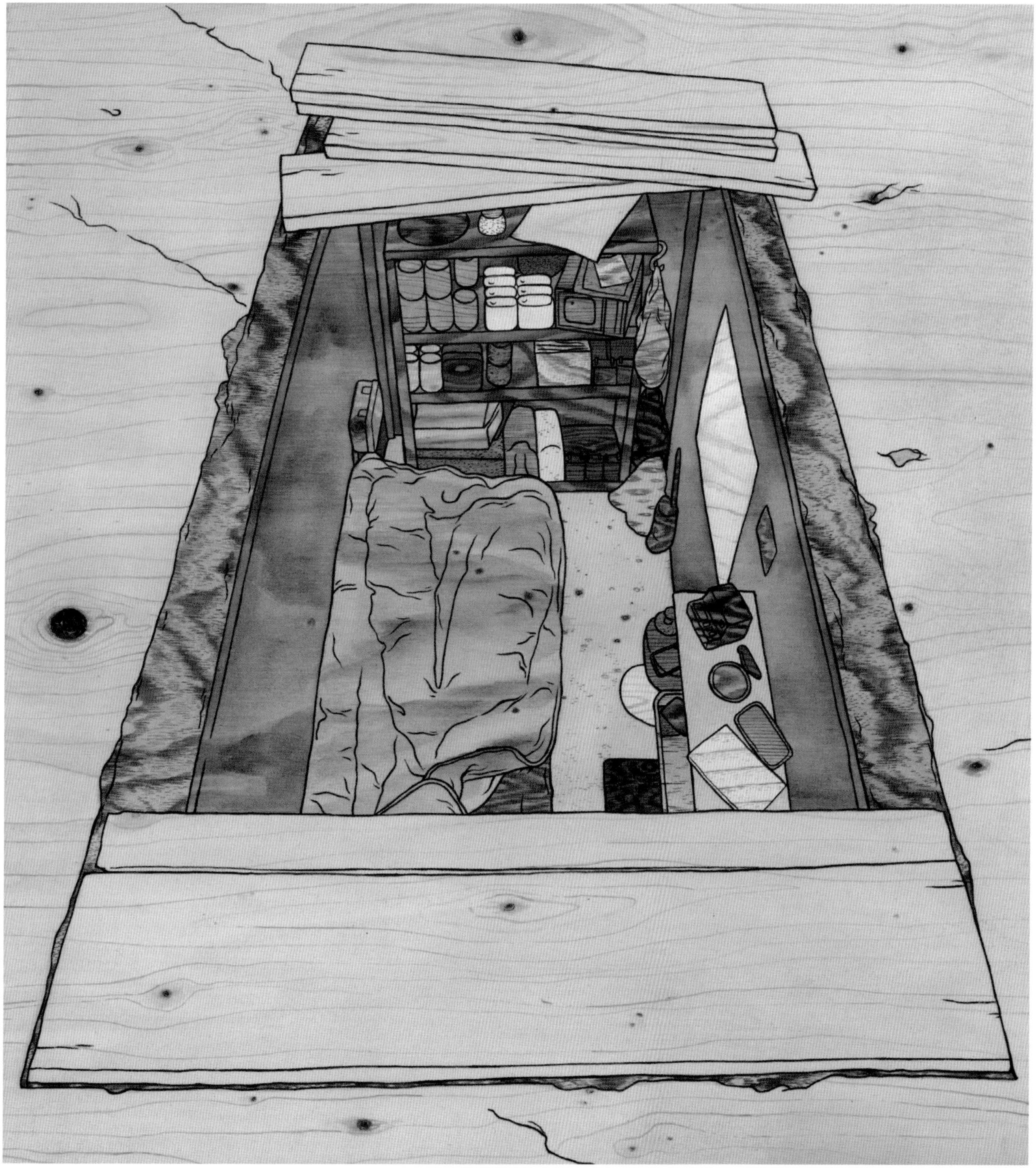

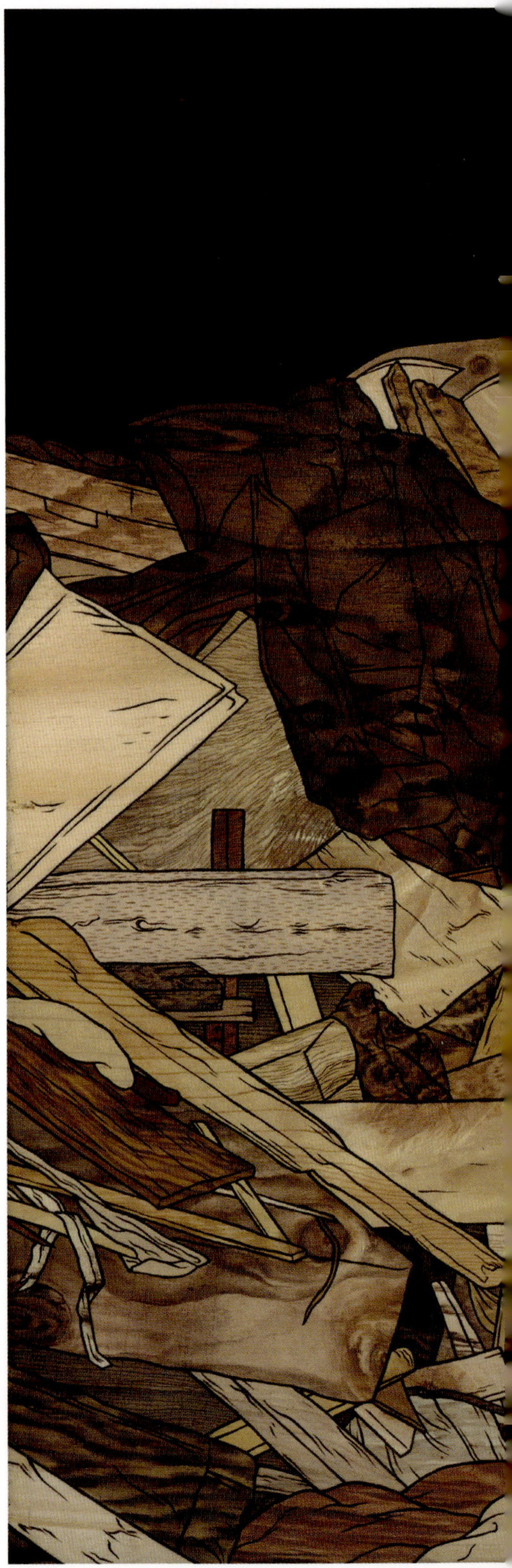

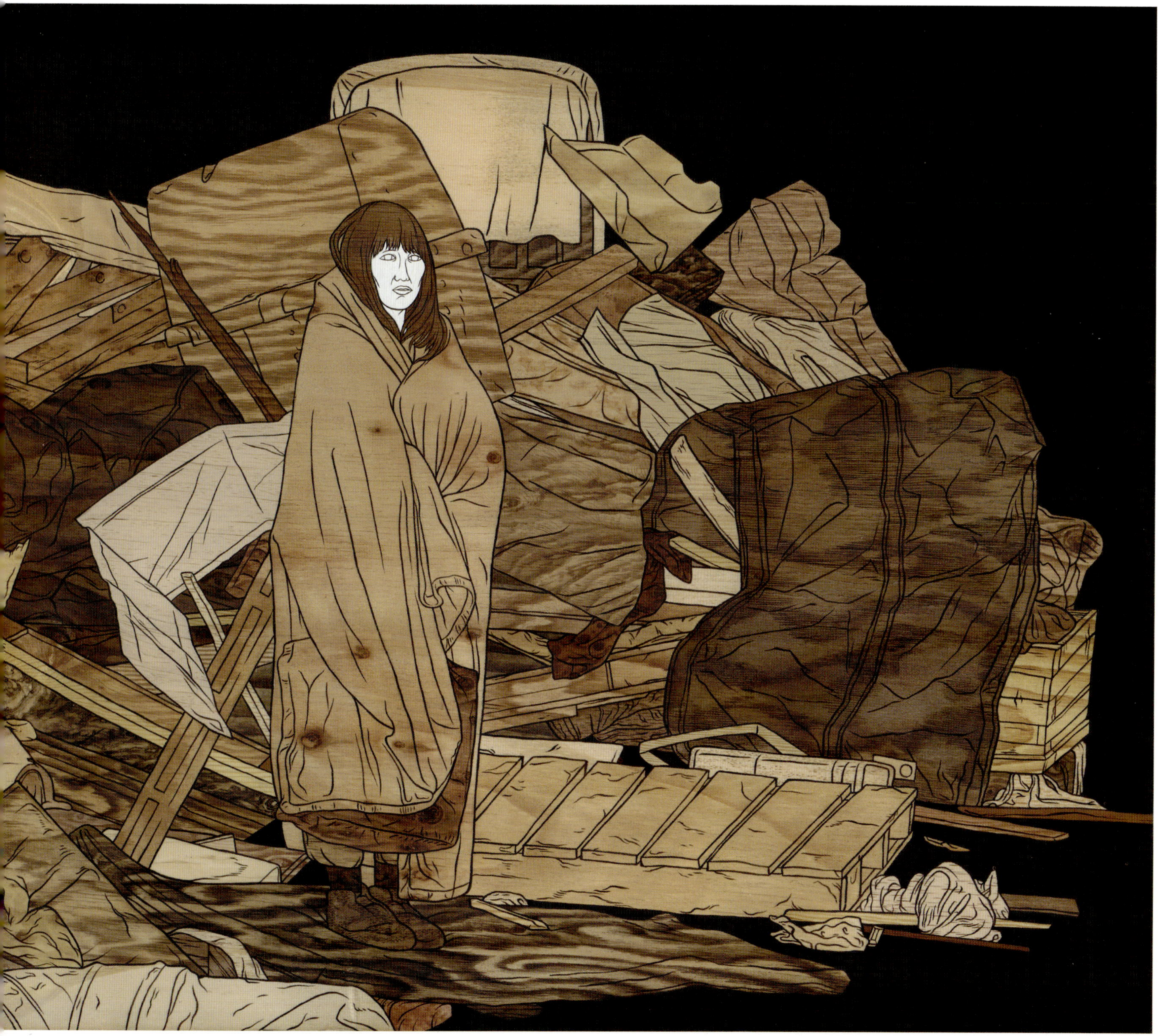

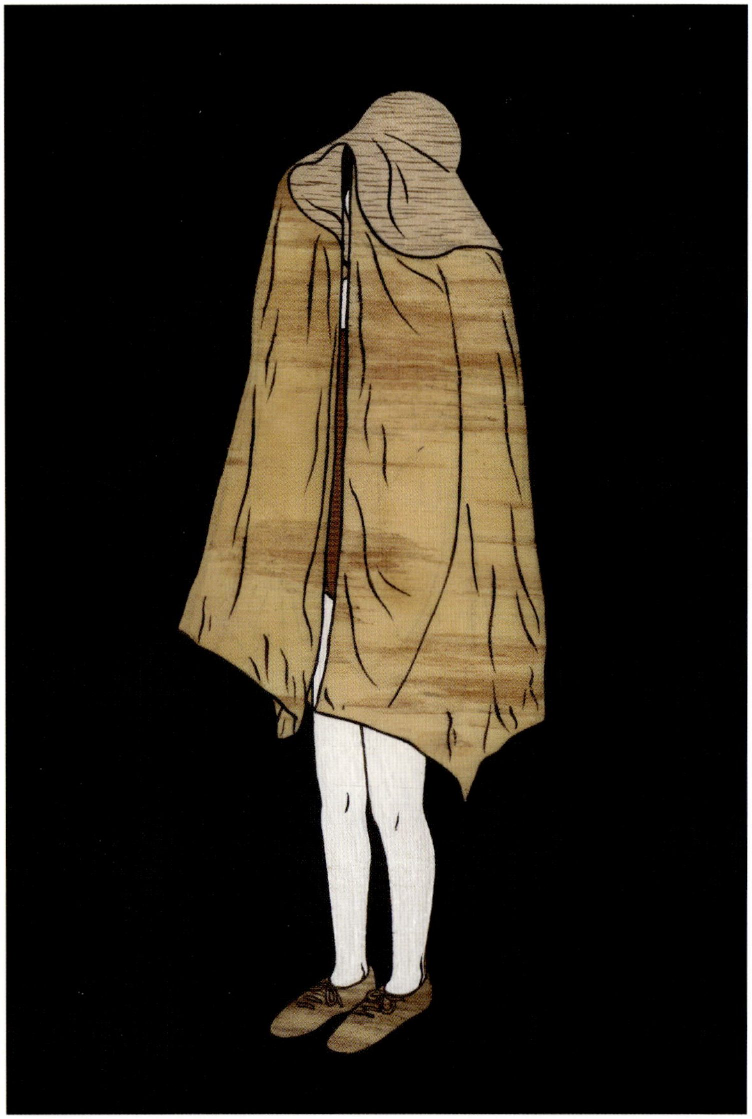

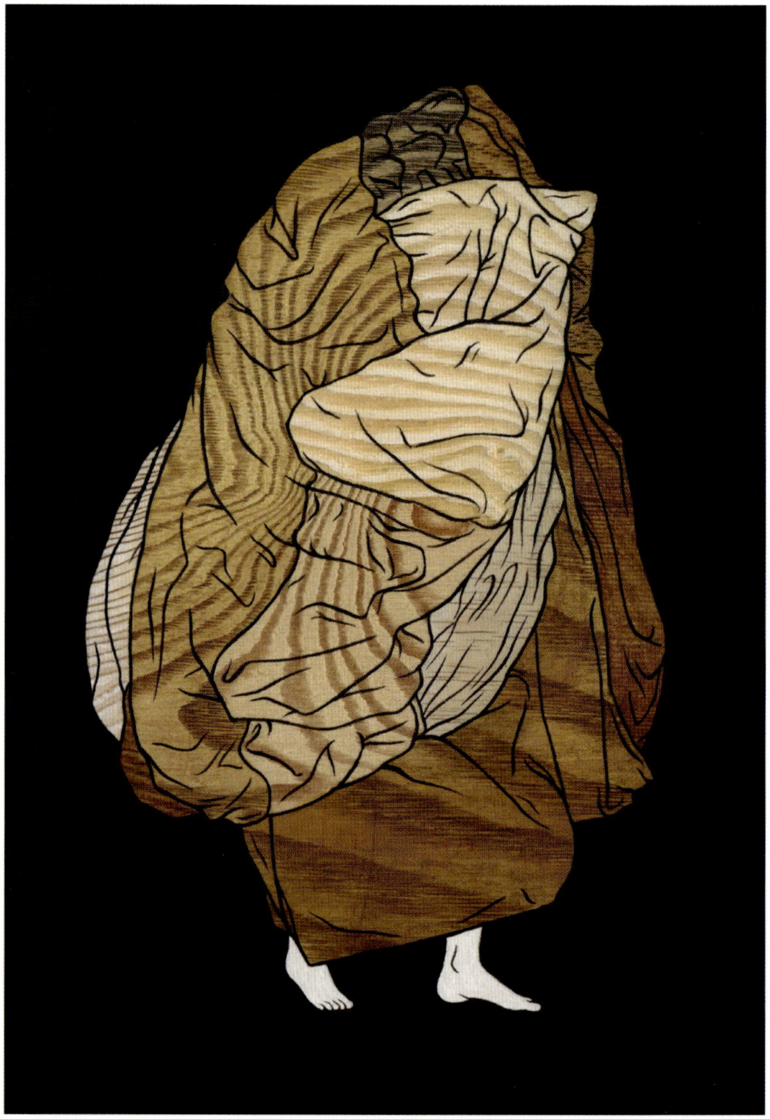

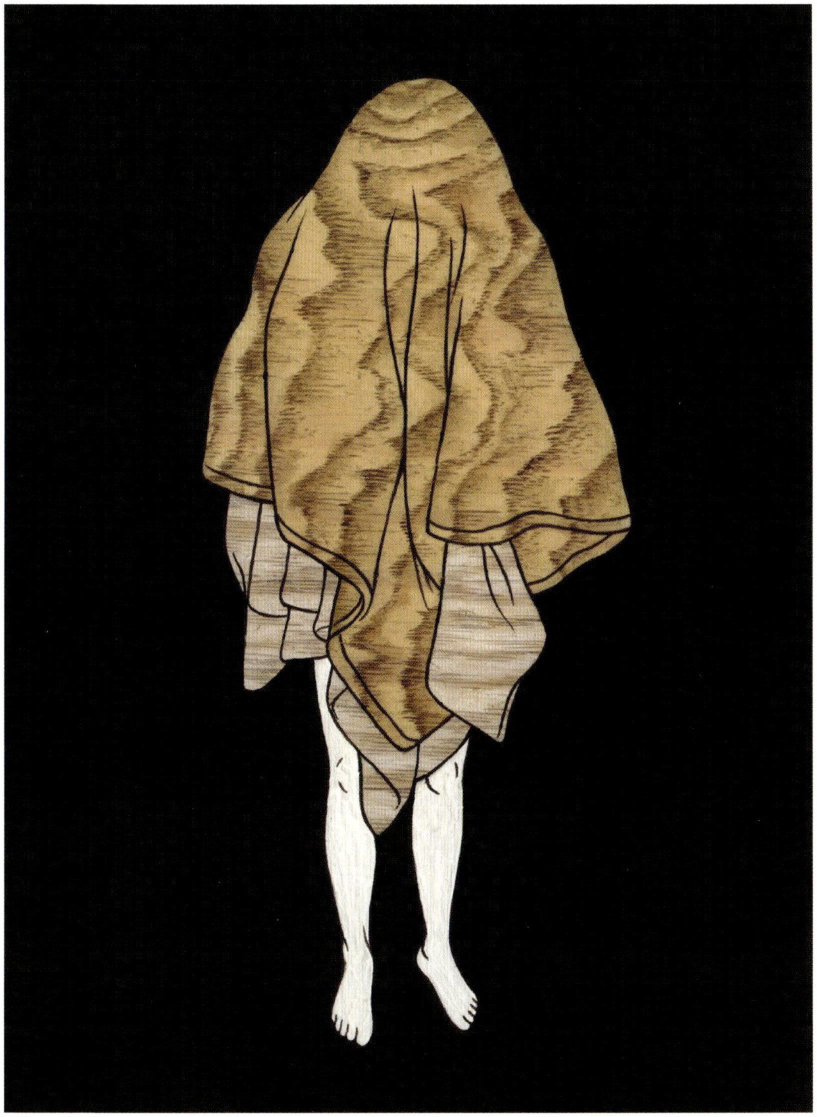

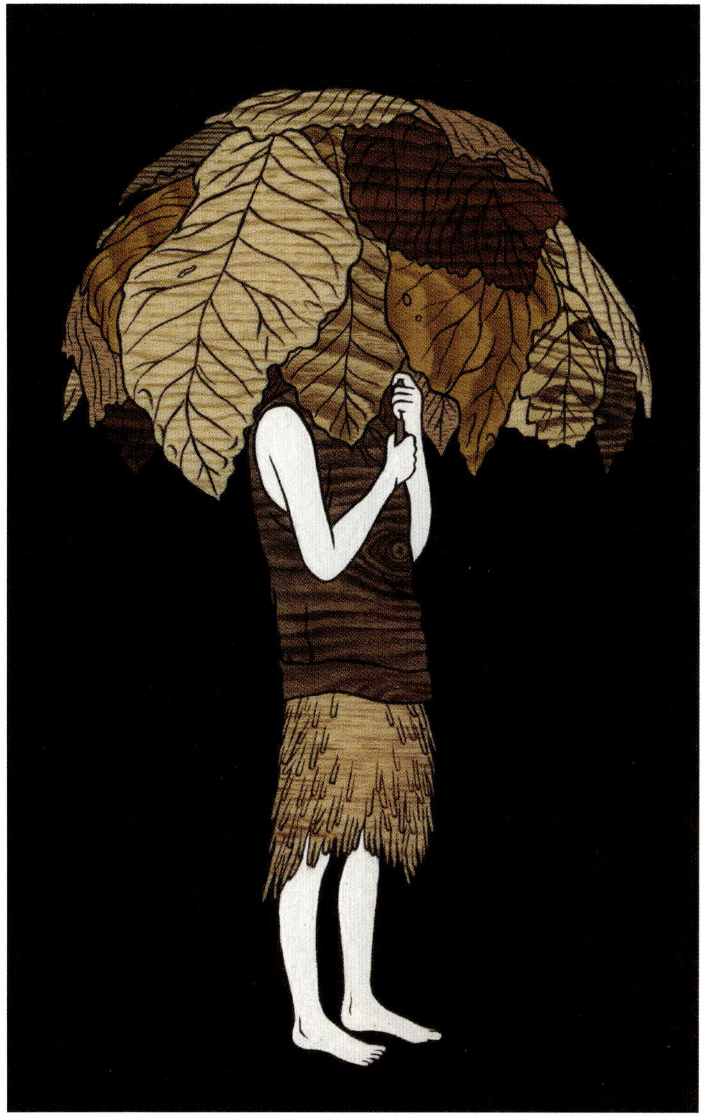

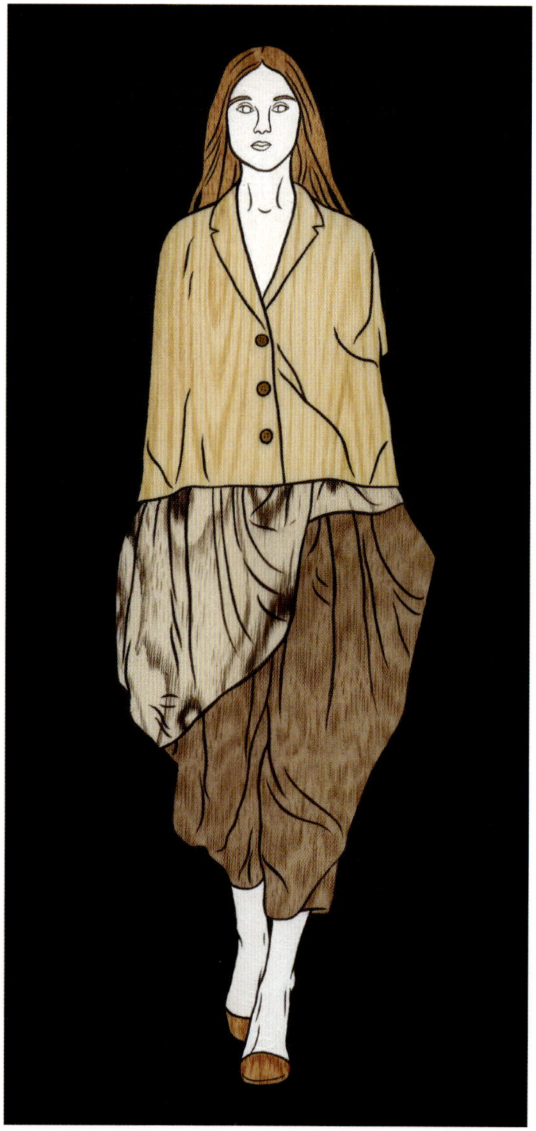

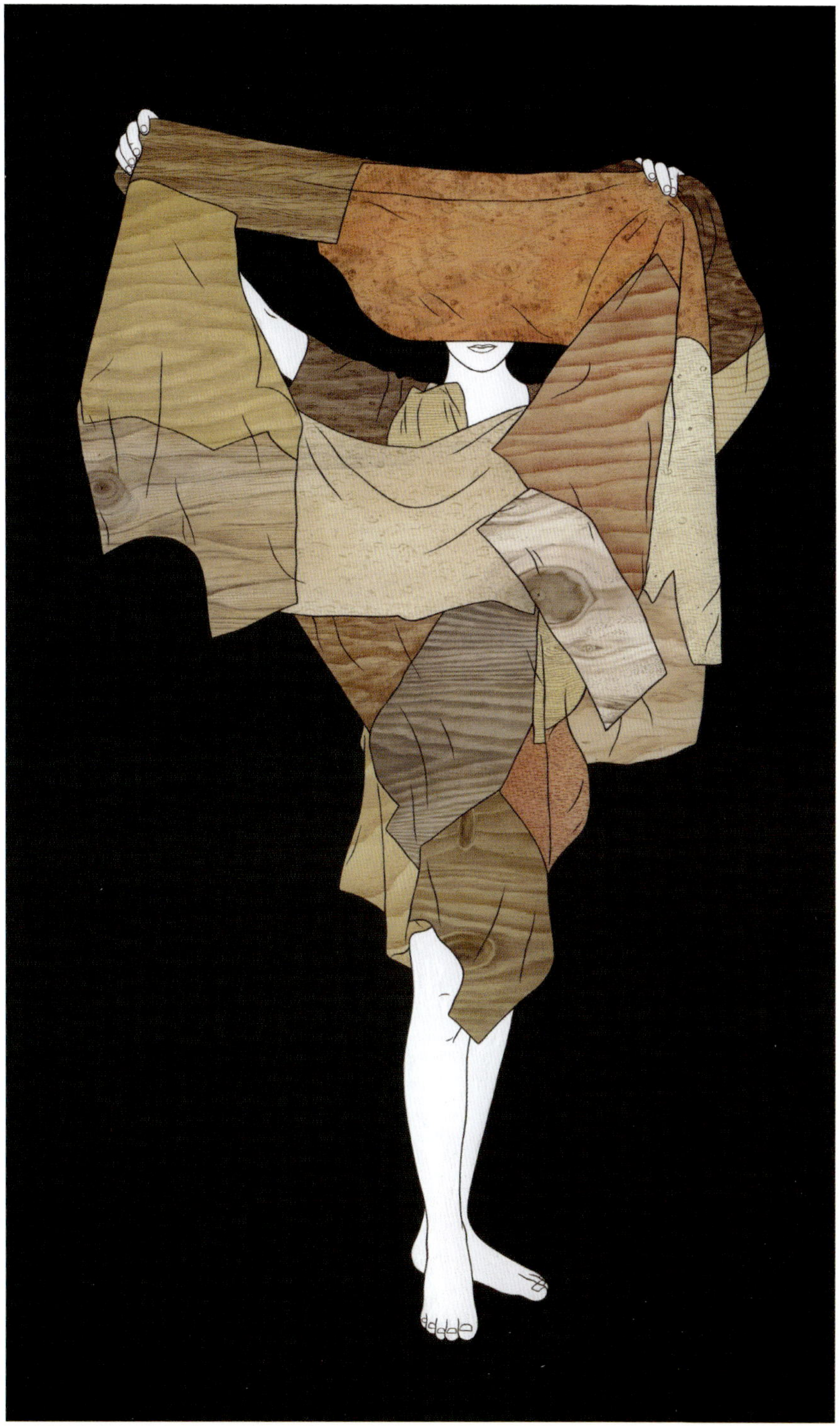

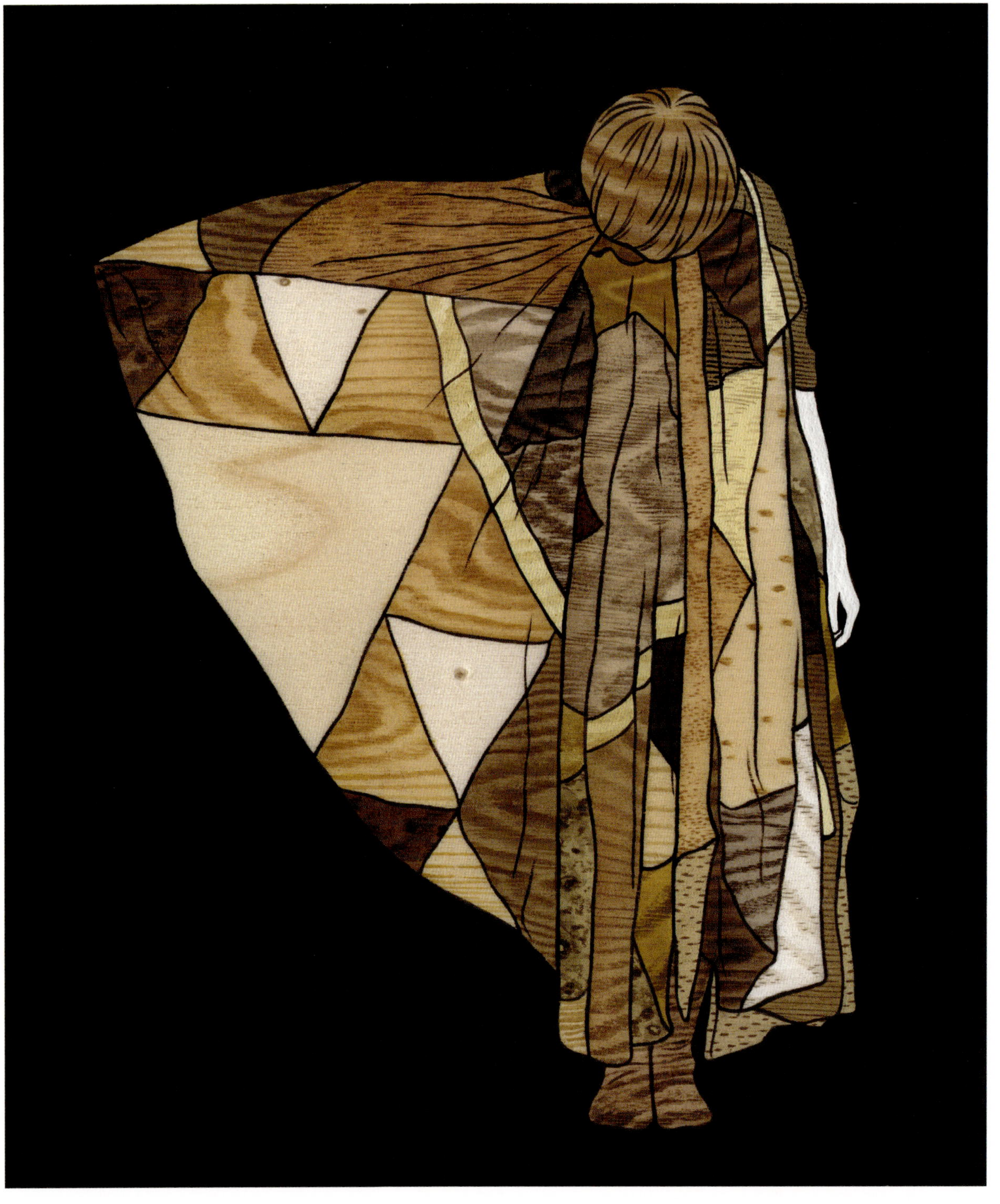

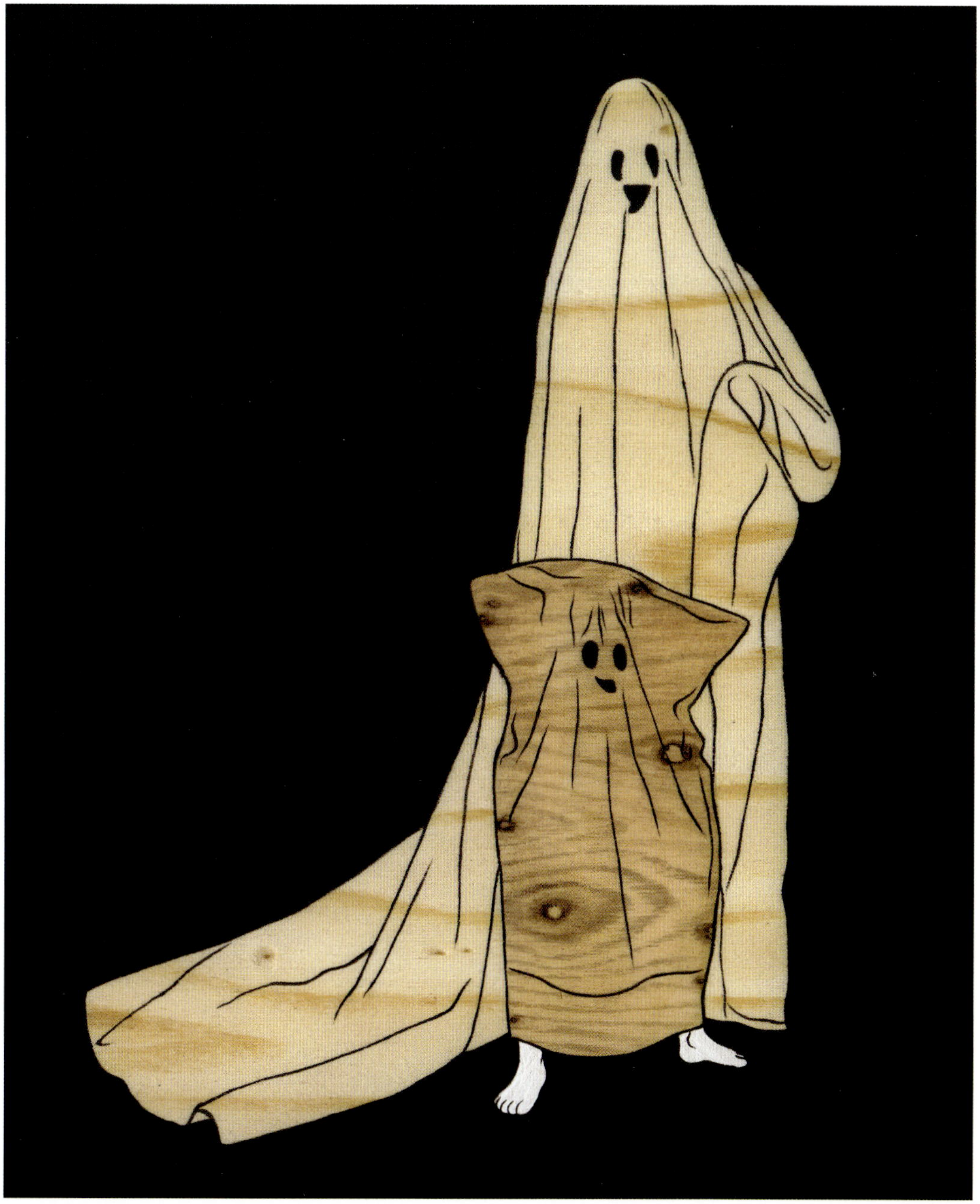

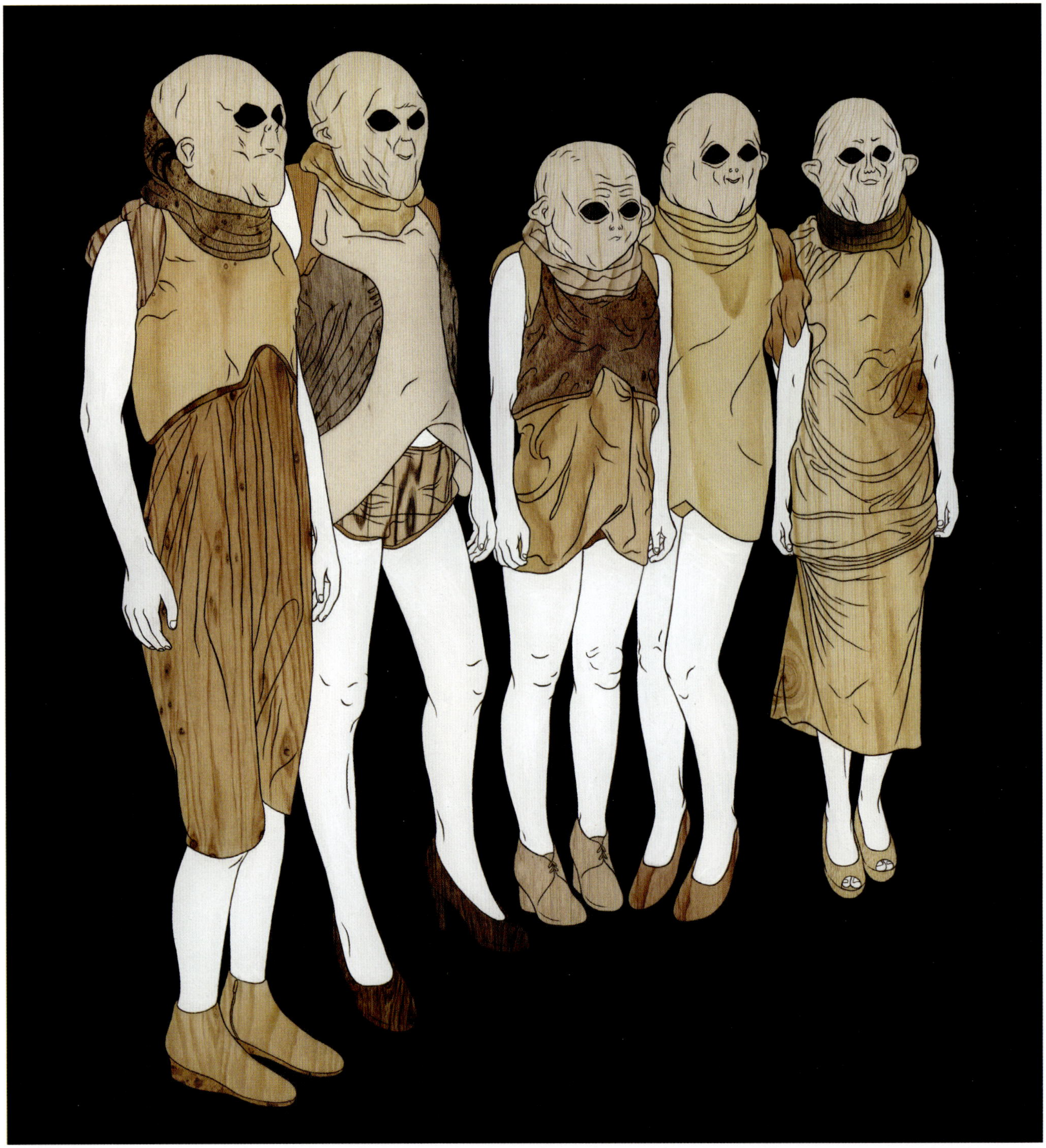

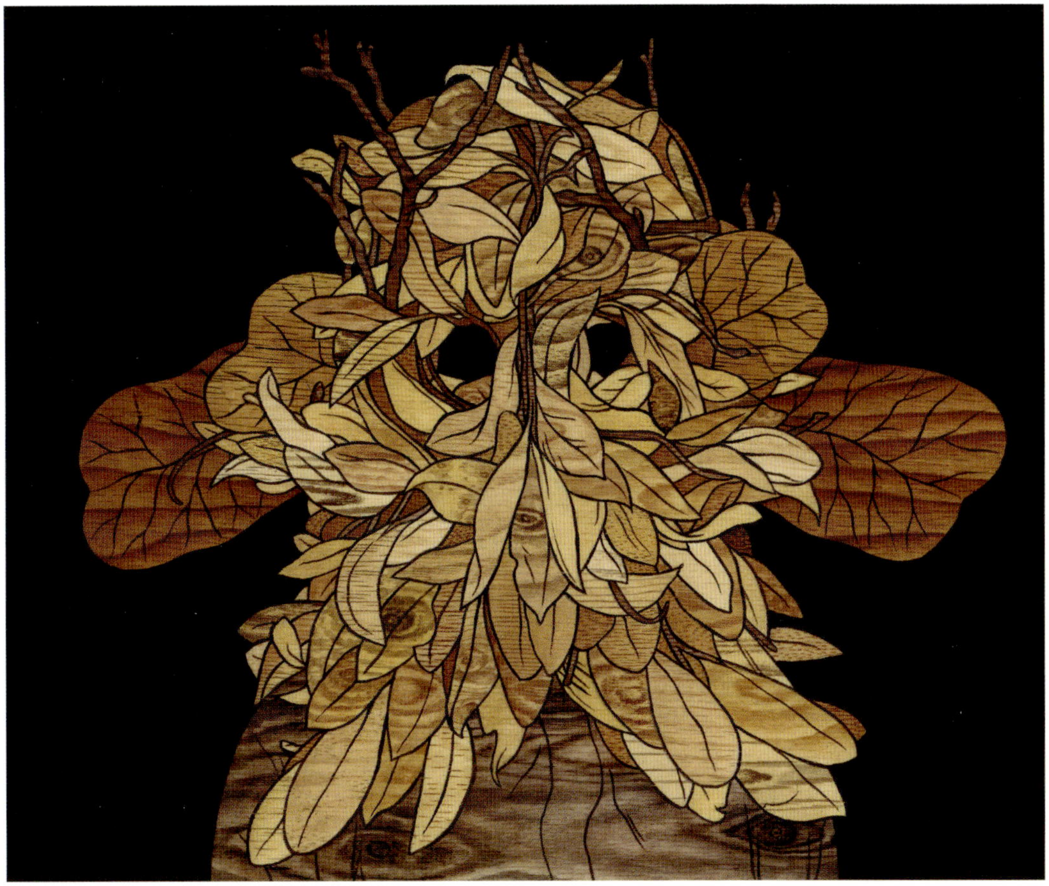

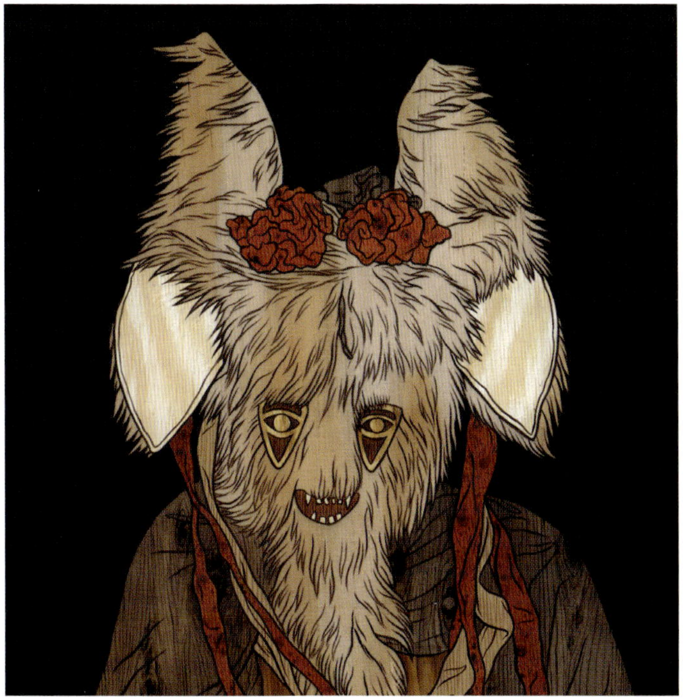

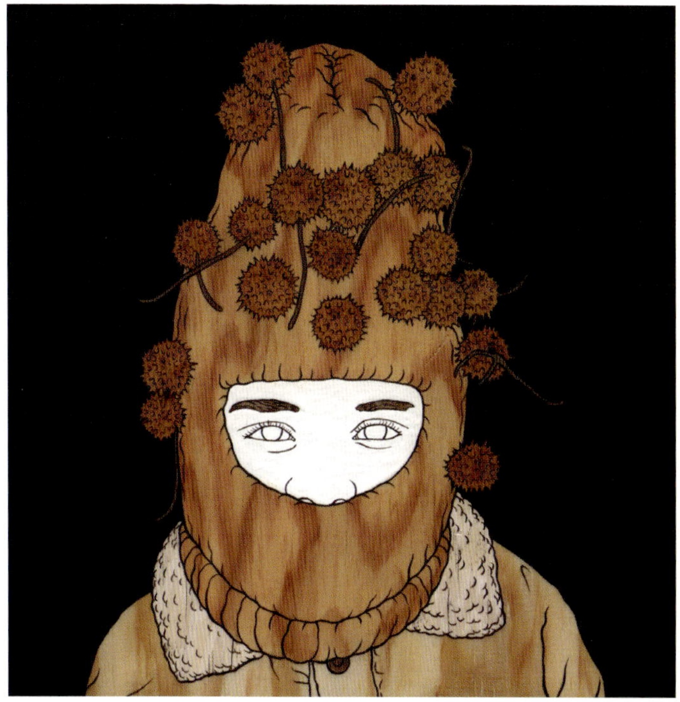

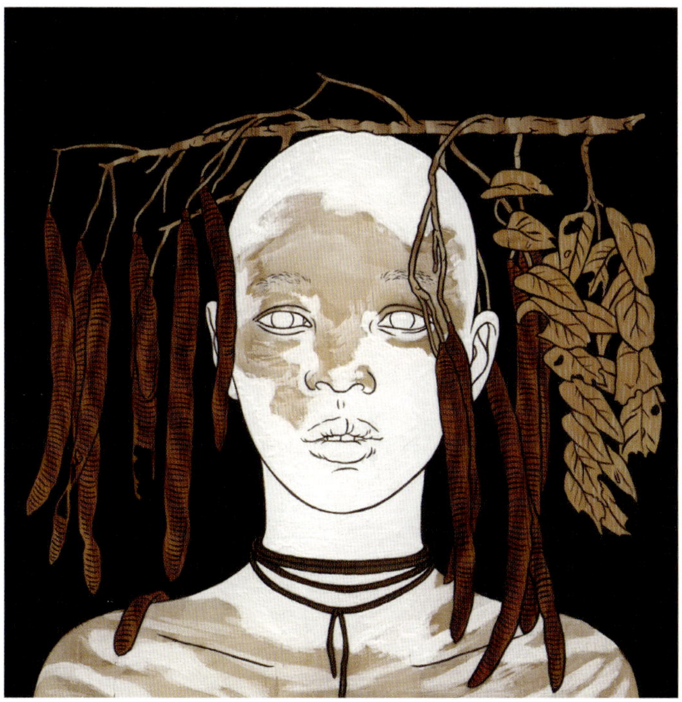

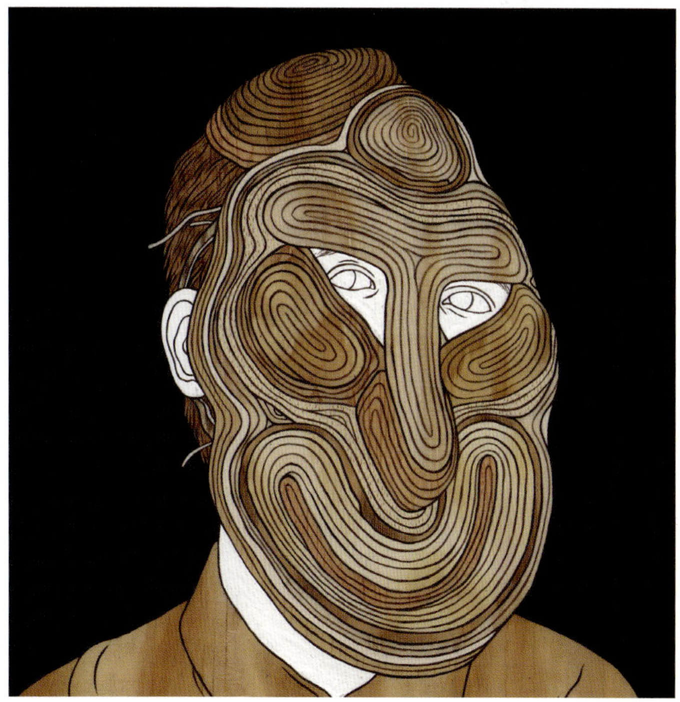

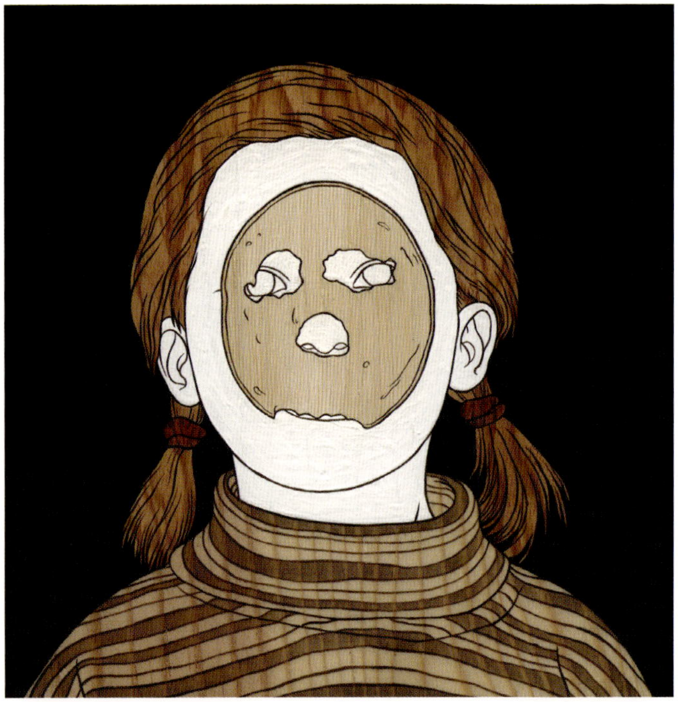

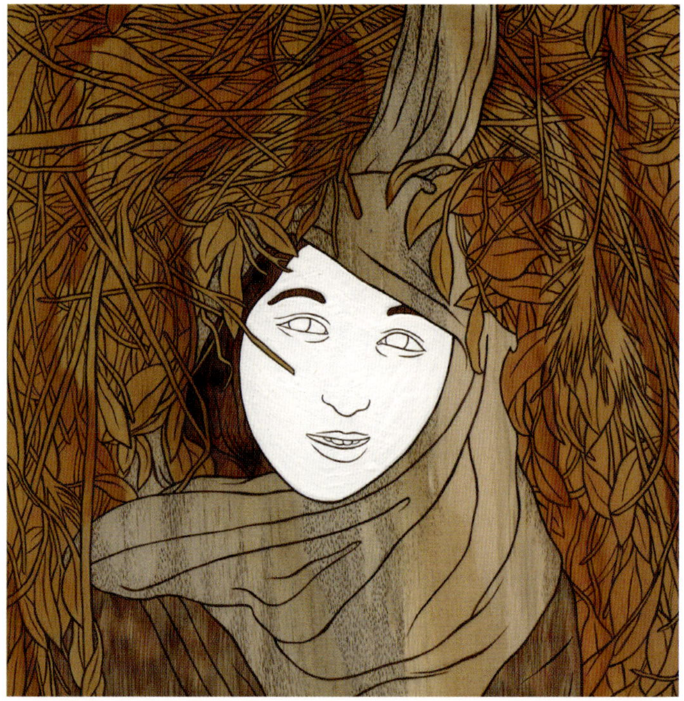

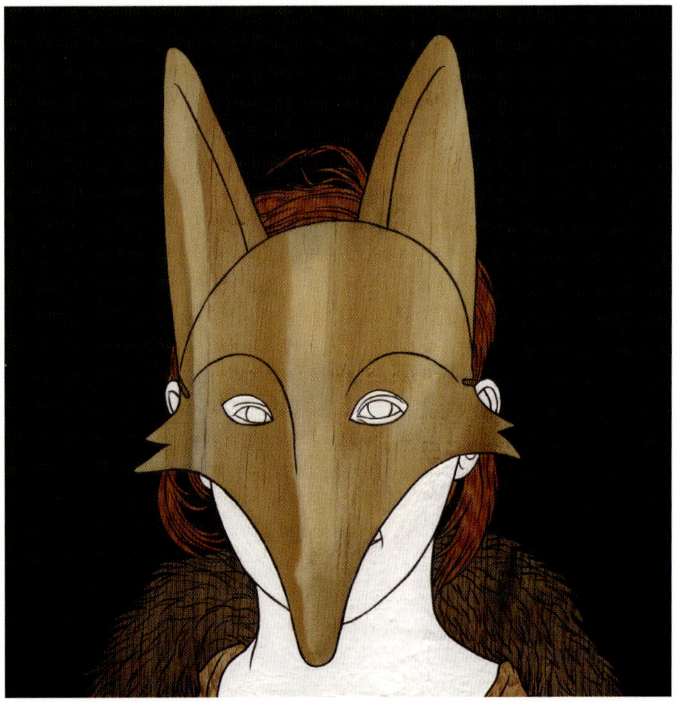

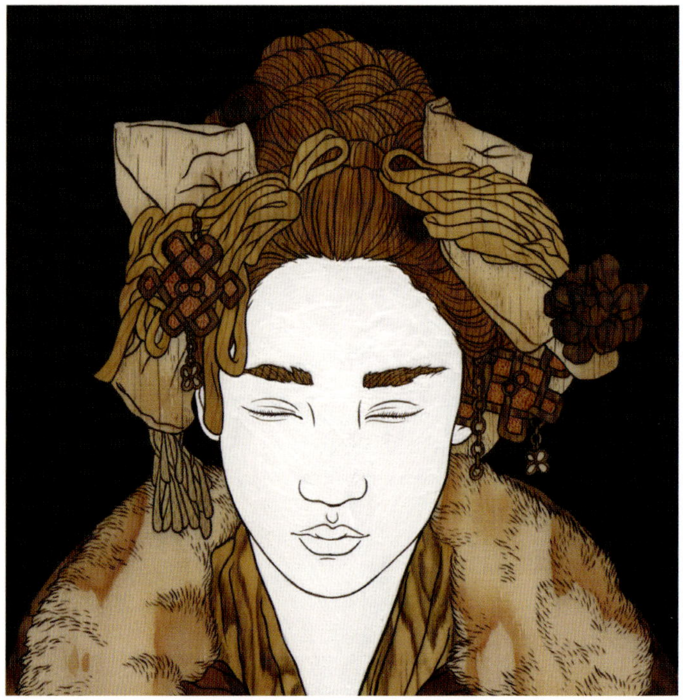

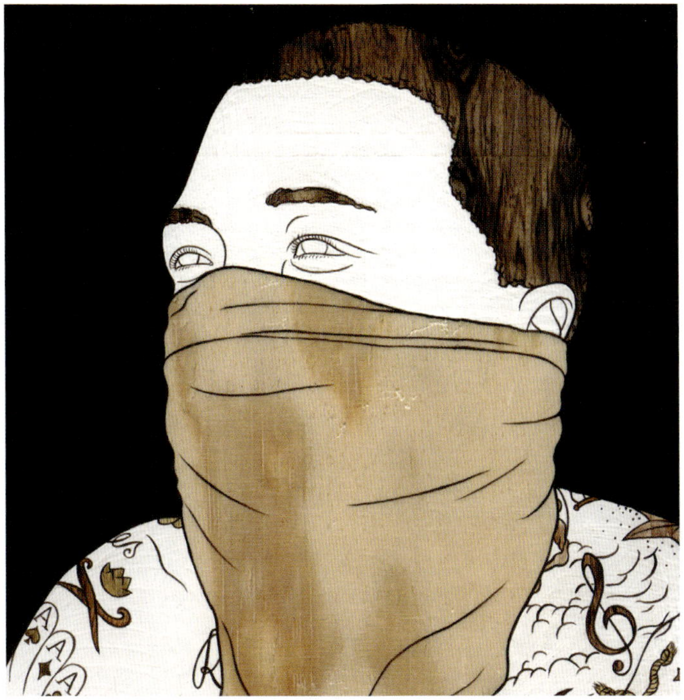

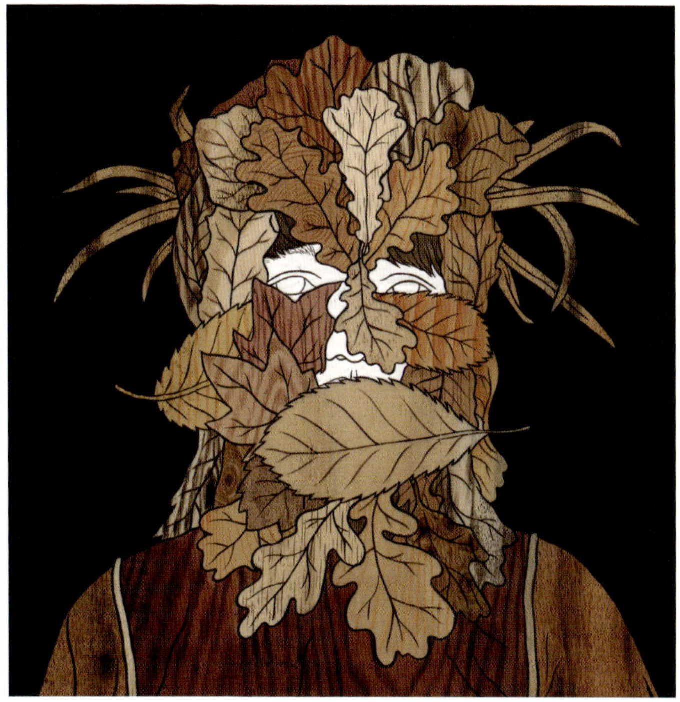

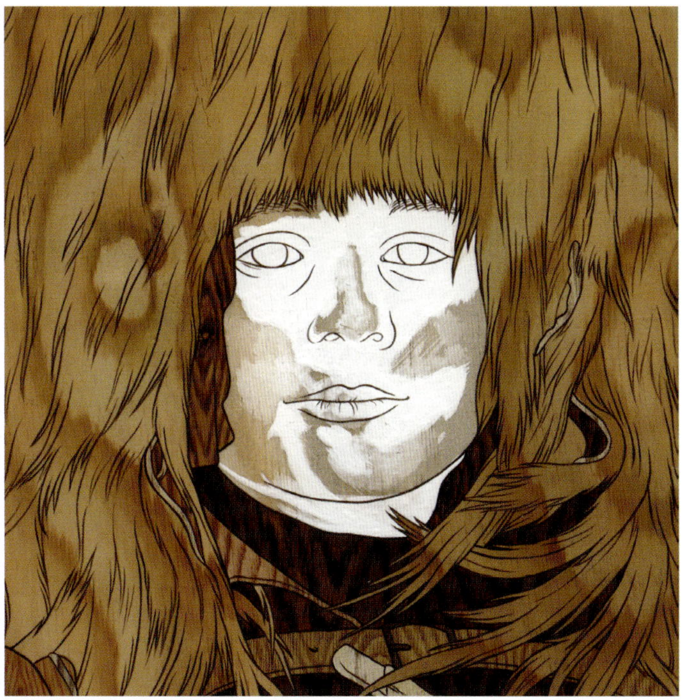

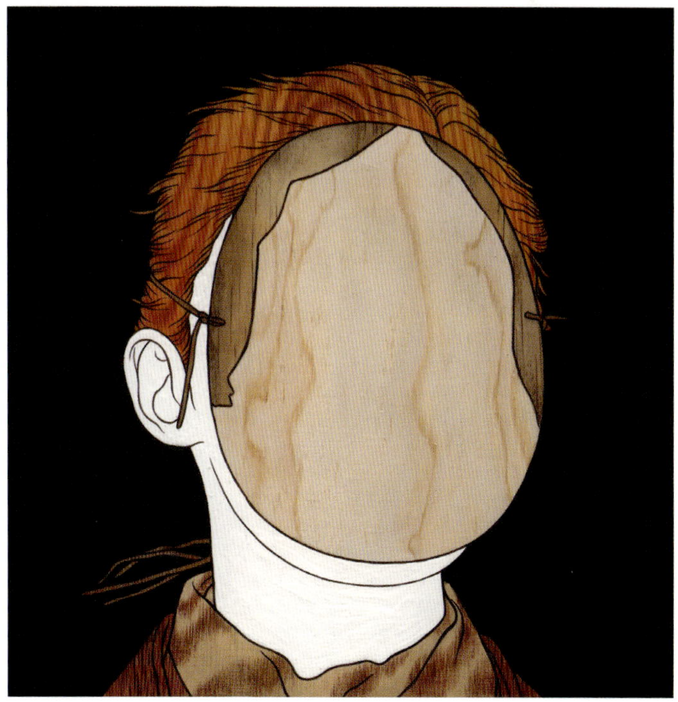

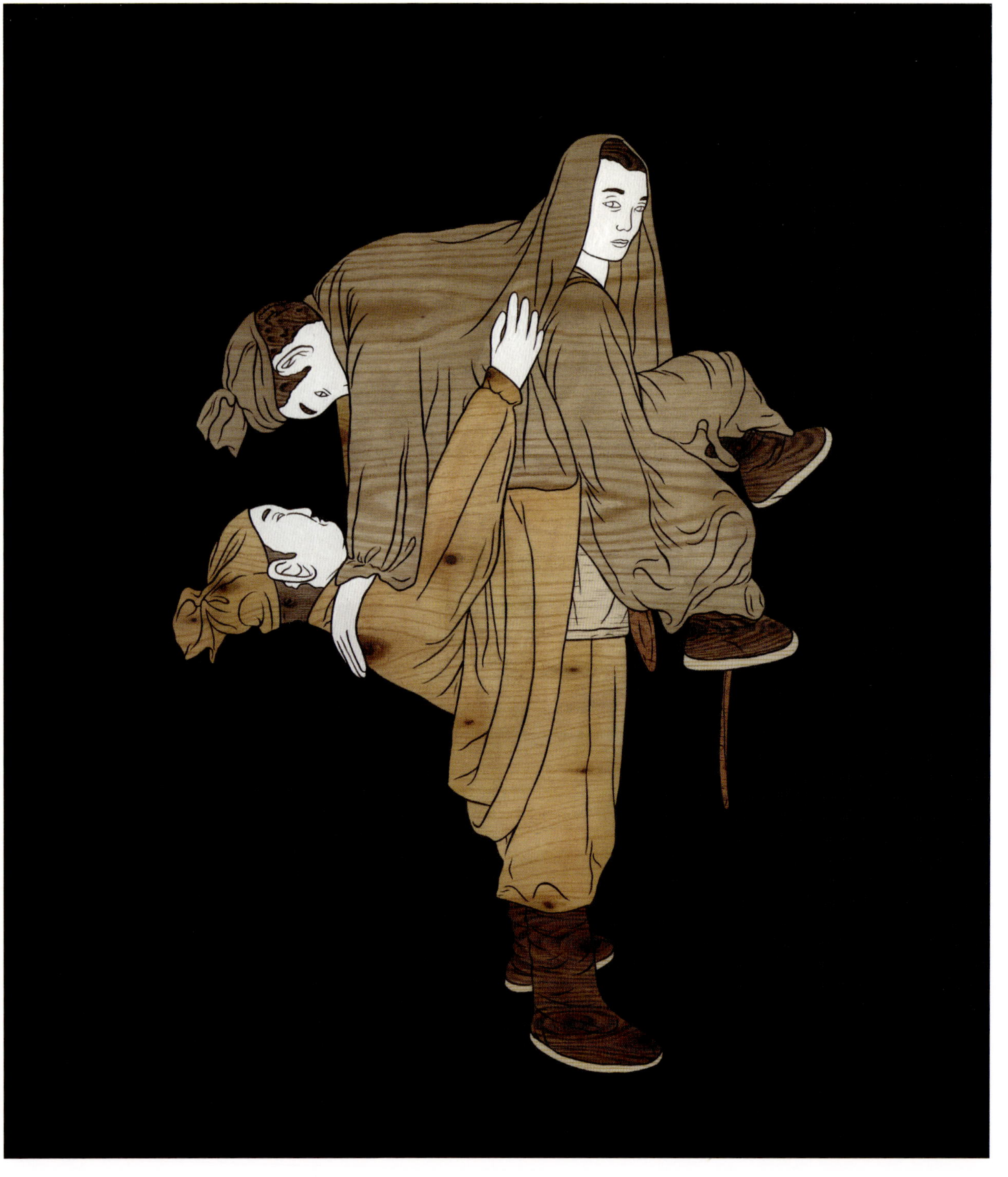

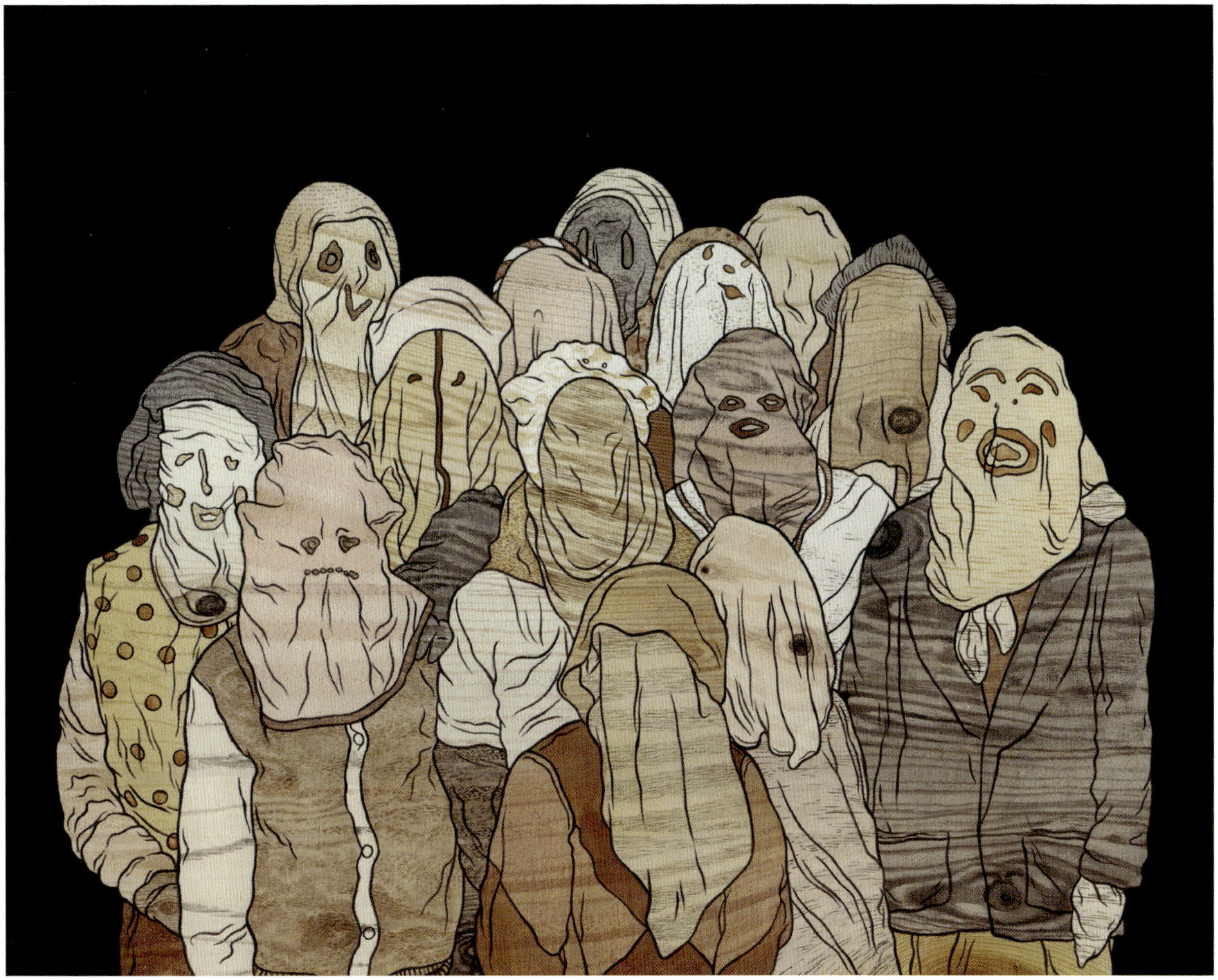

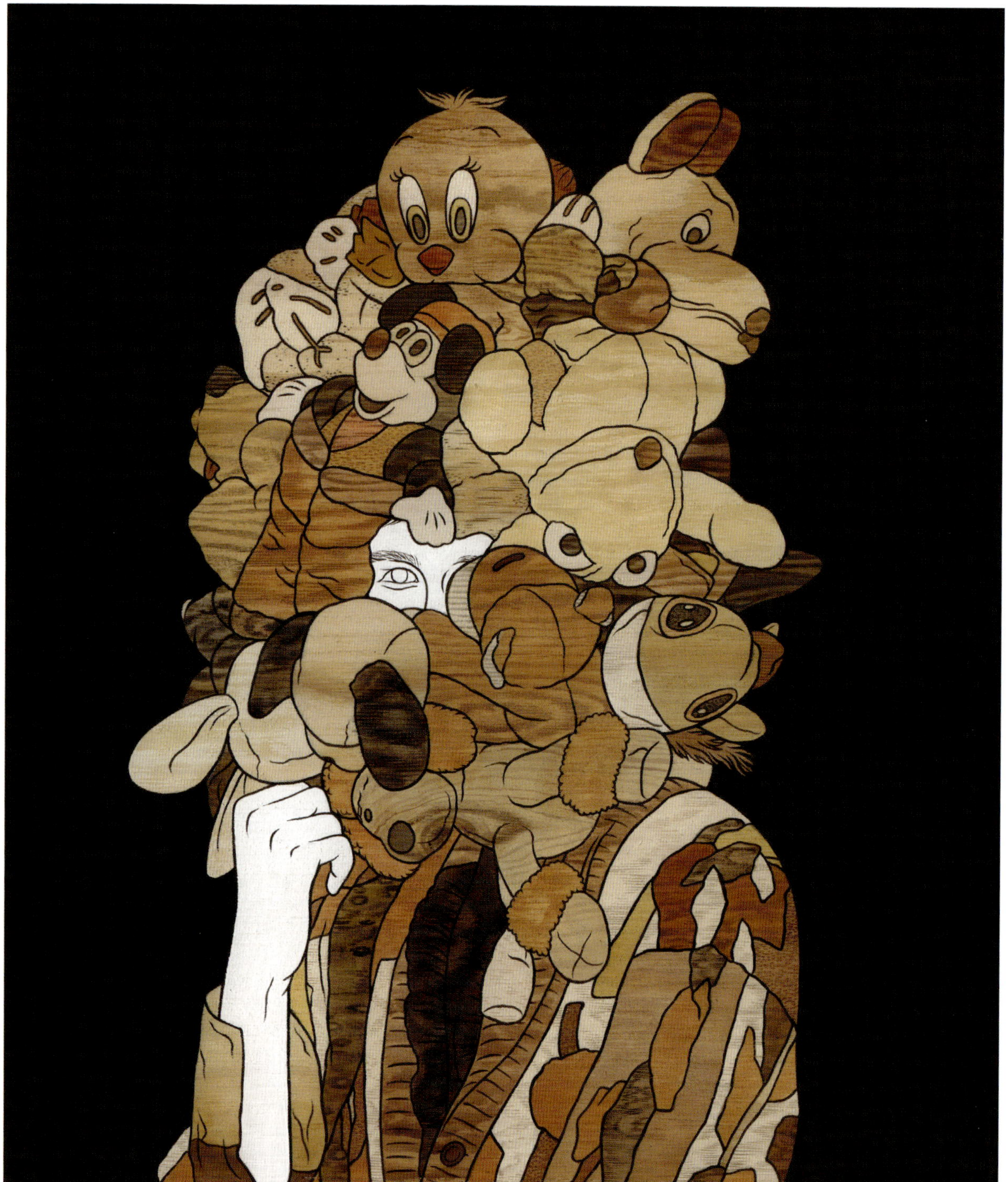

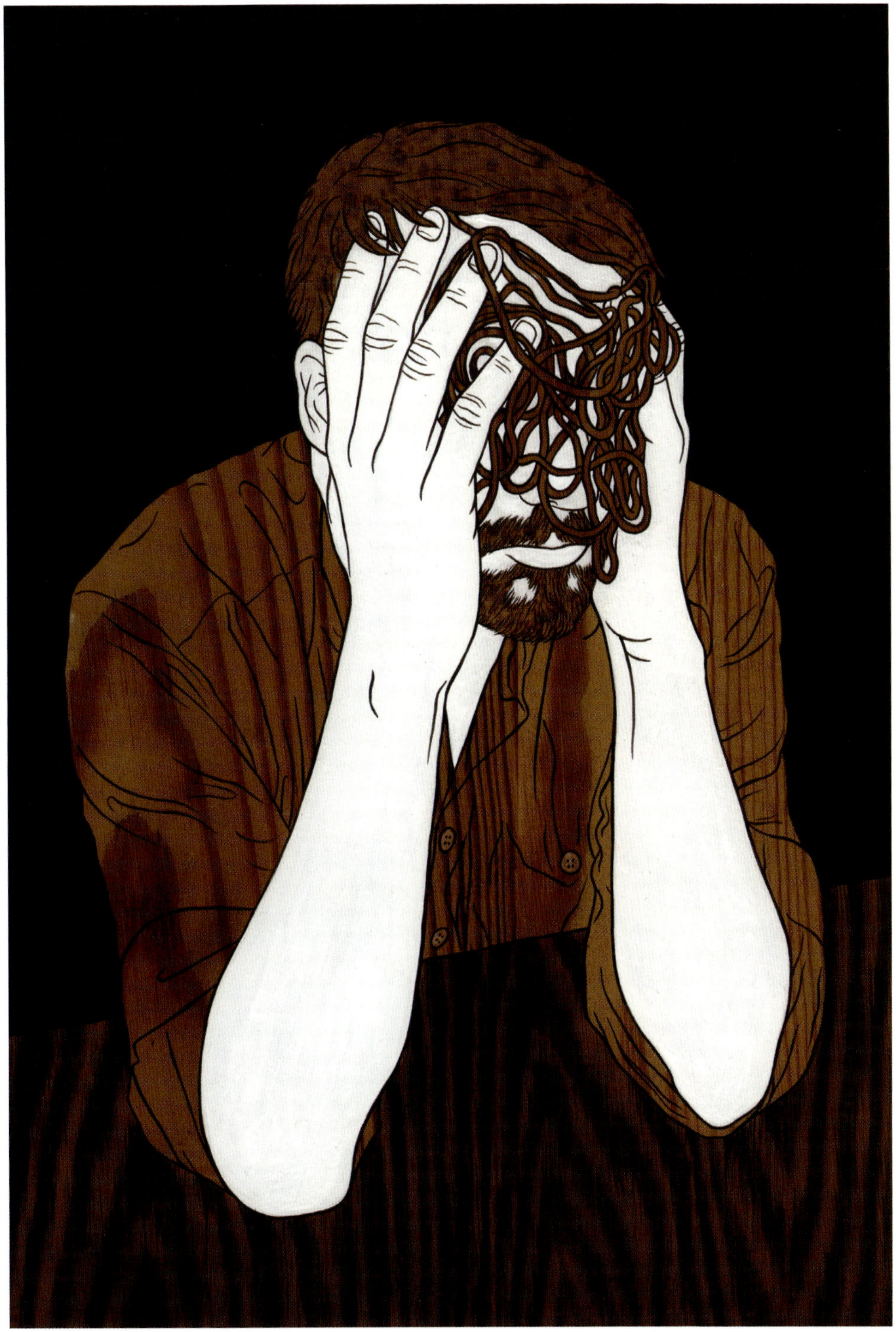

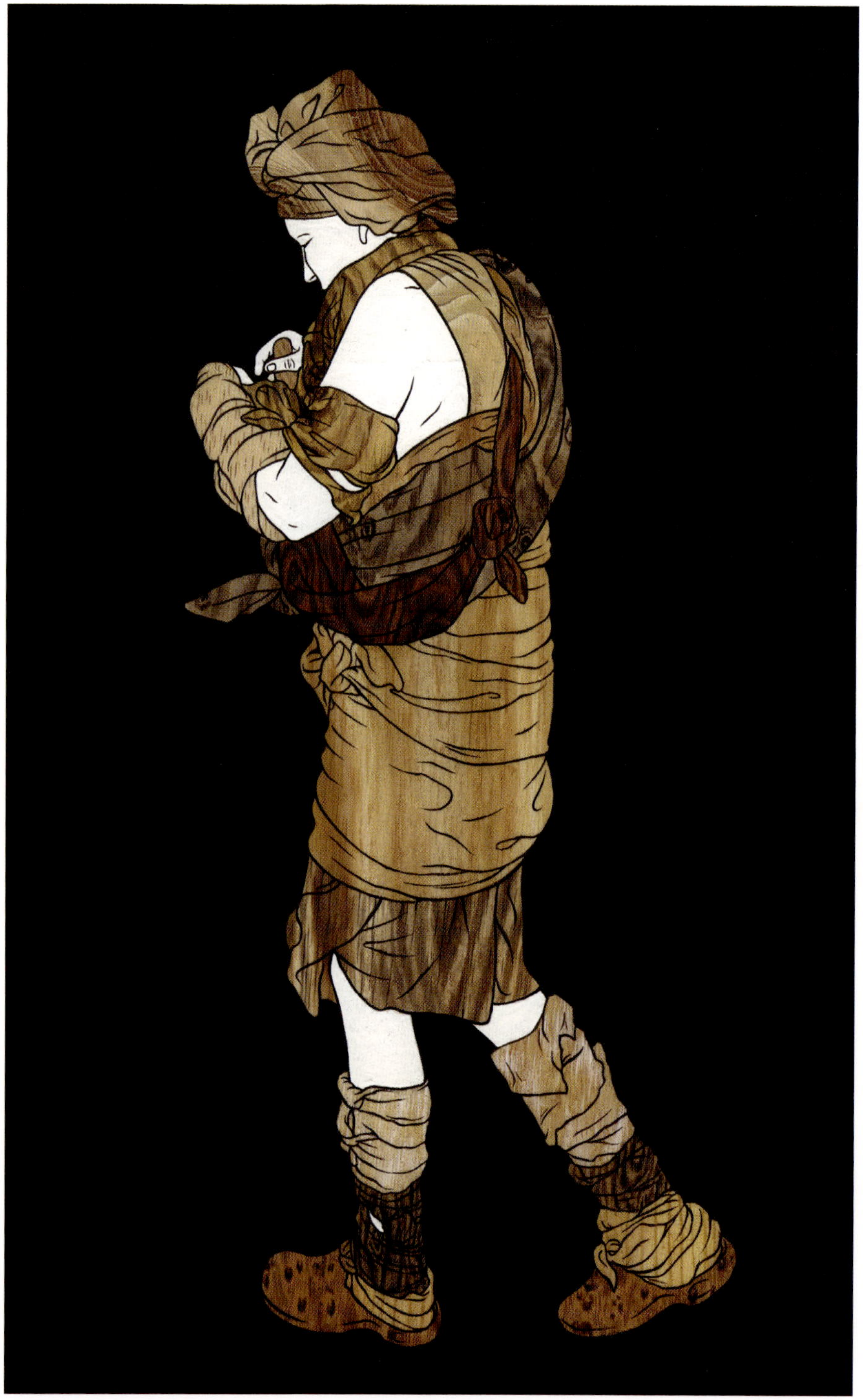

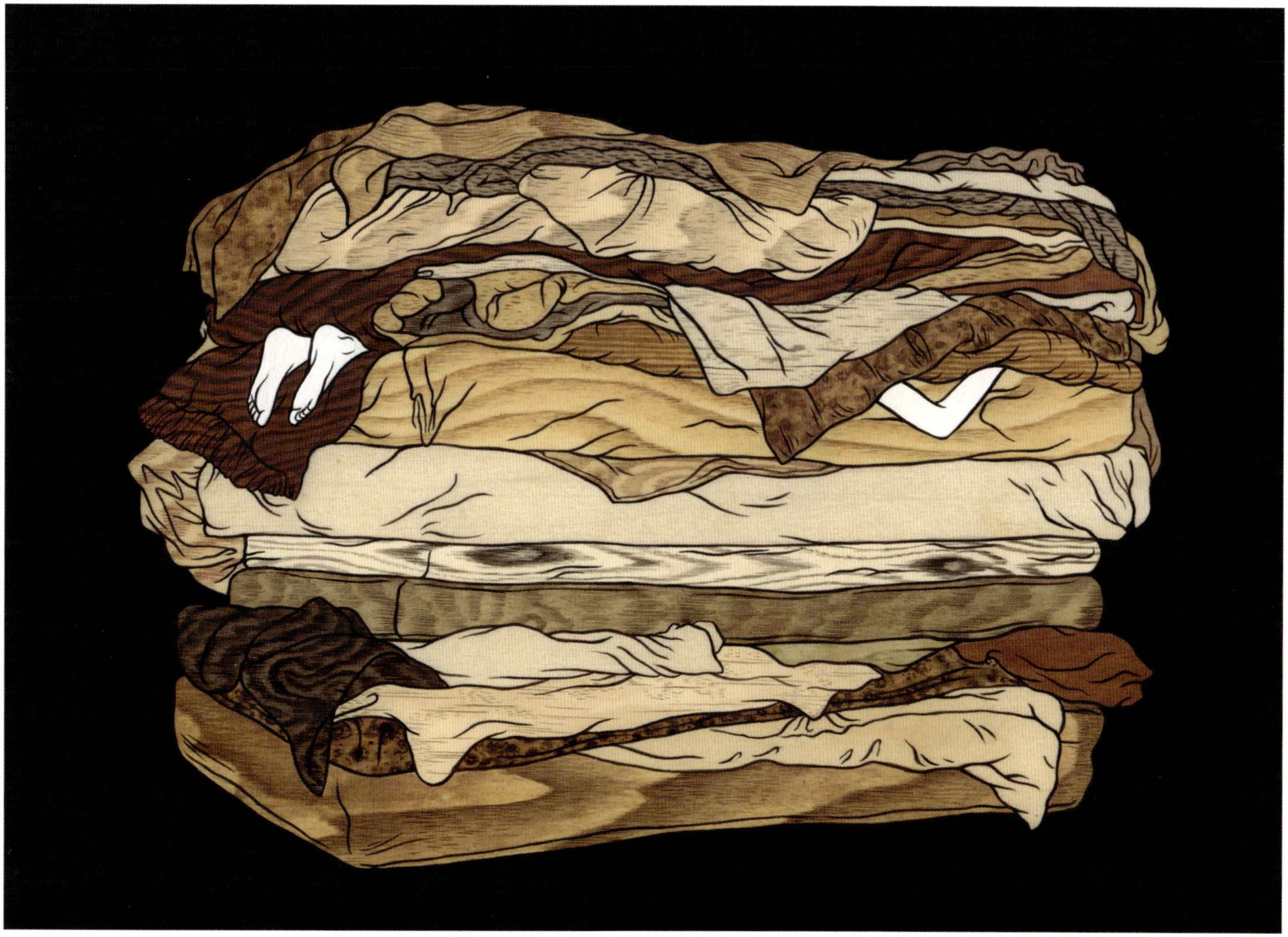

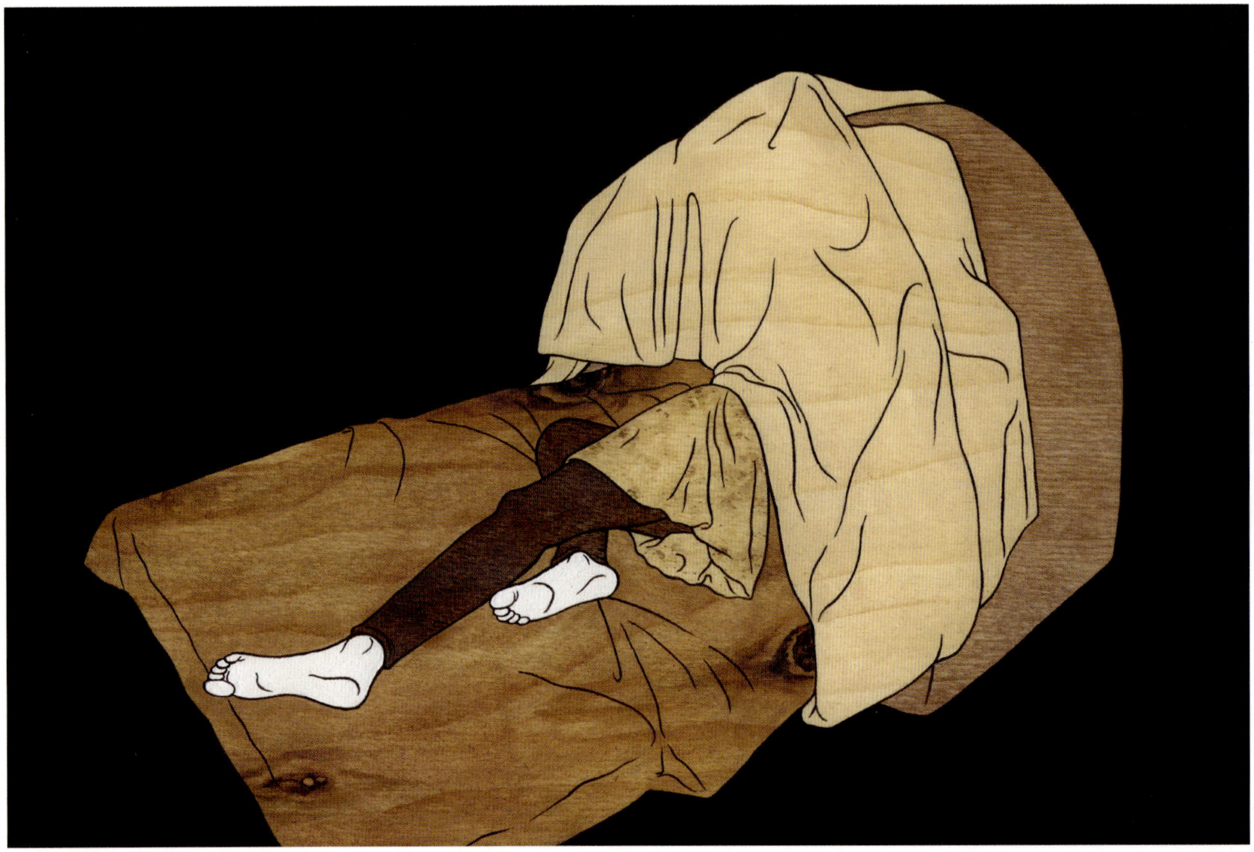

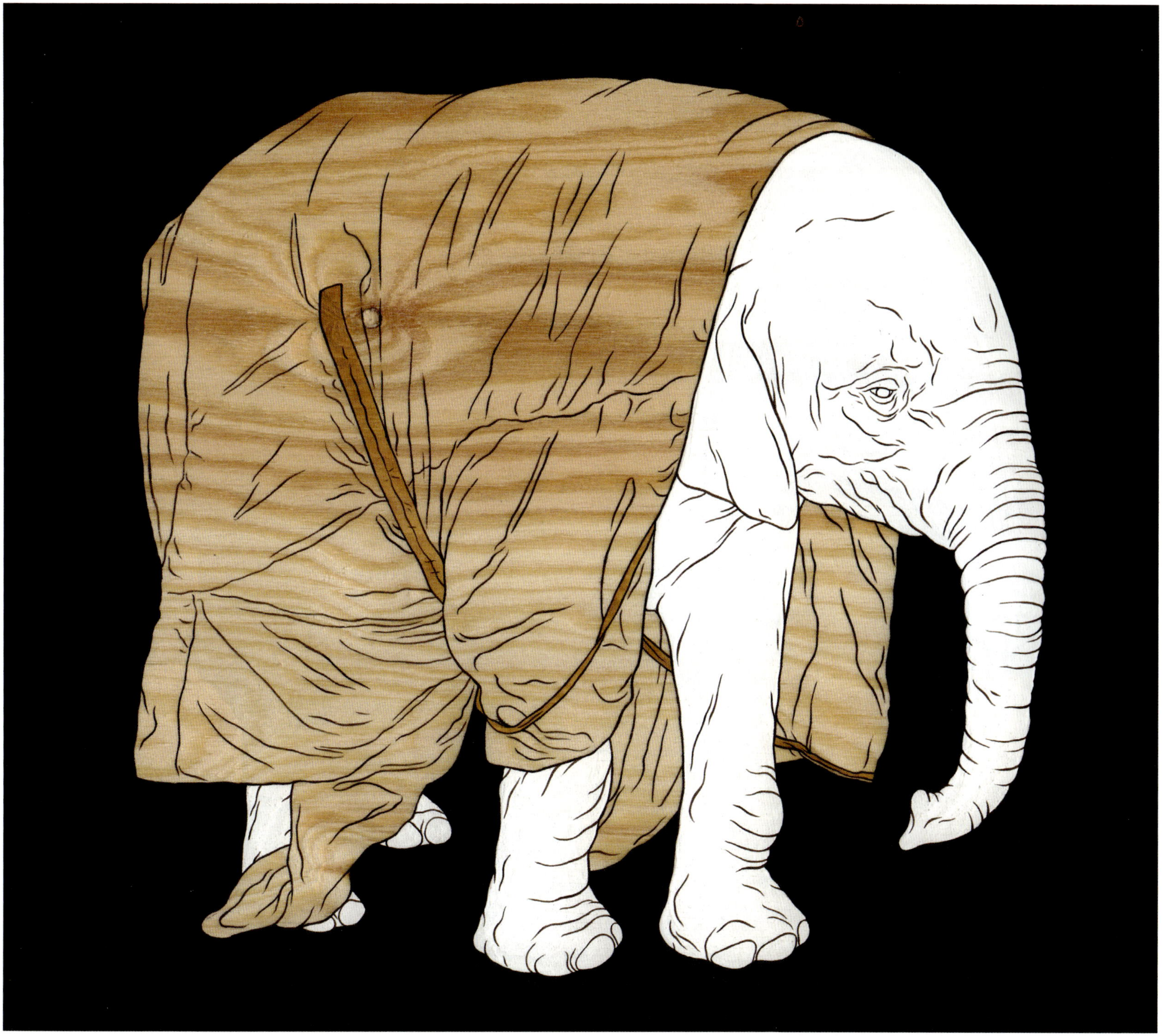

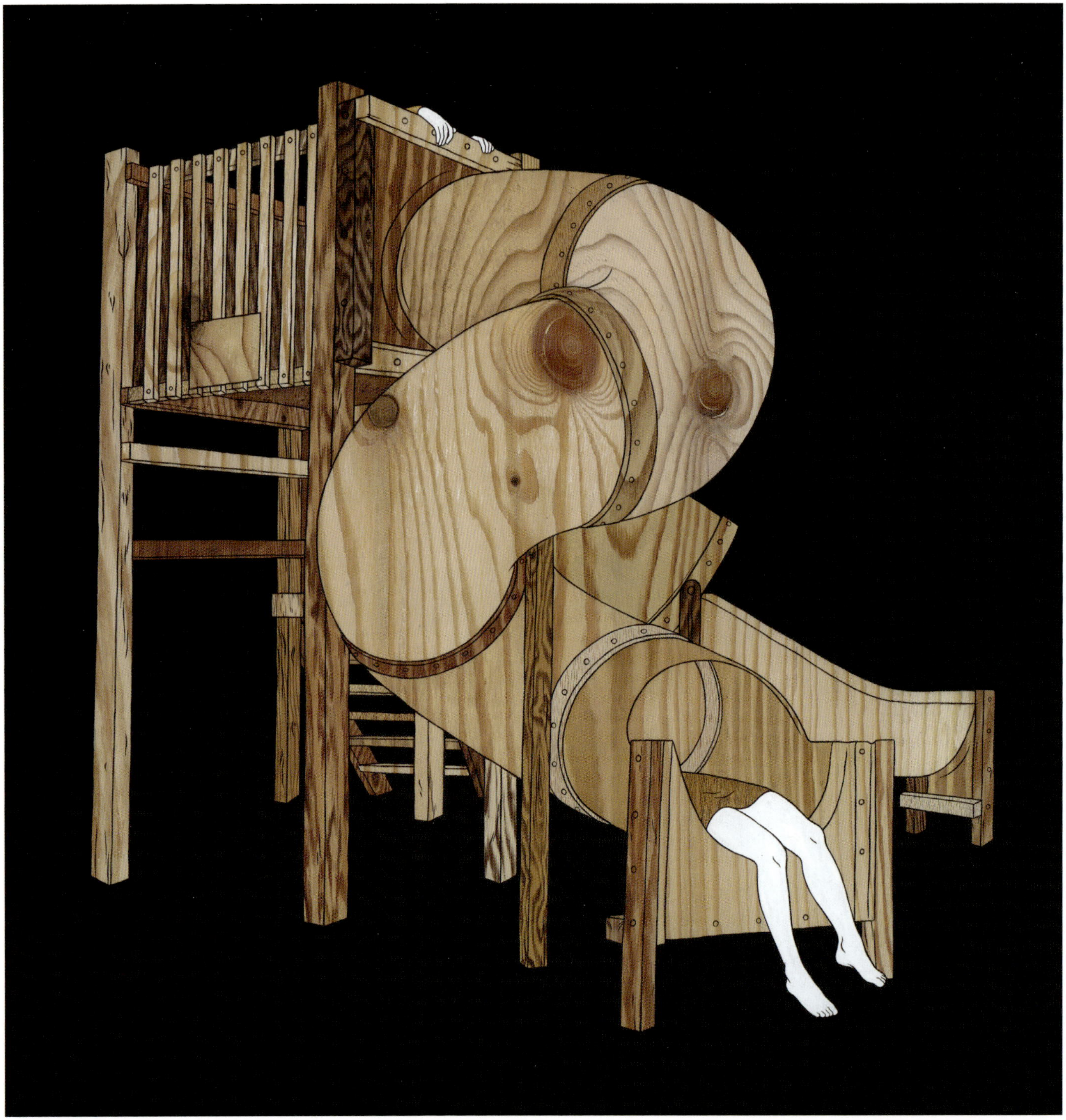

Plates Index

Acrylic on wood

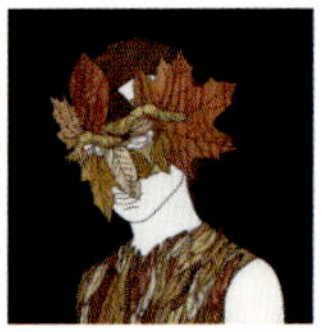

p 11 25,5x25,5 cm | 2012

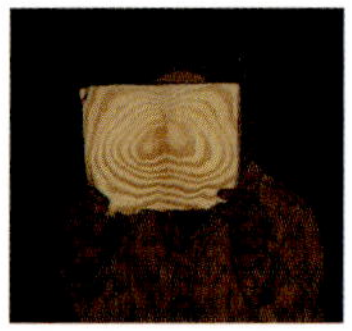

p 12 19x21 cm | 2010

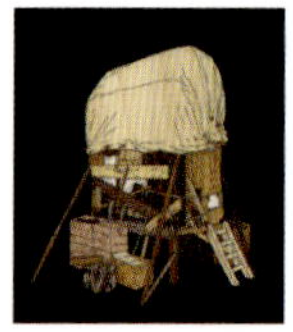

p 13 47x42 cm | 2012

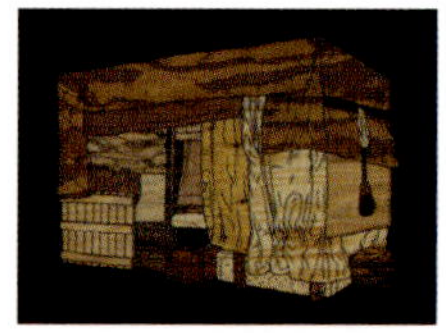

p 14 30x40 cm | 2010

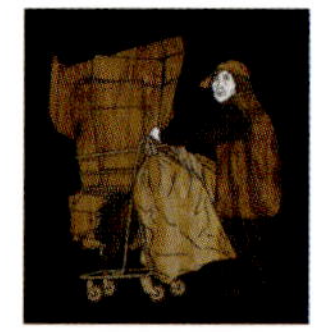

p 15 30x28 cm | 2010

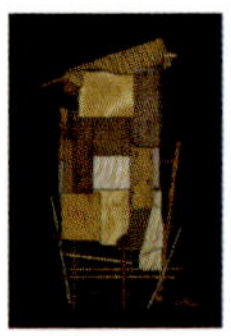

p 17 42x30 cm | 2010

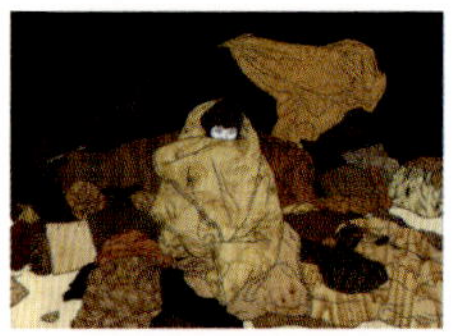

p 18 52x75 cm | 2010

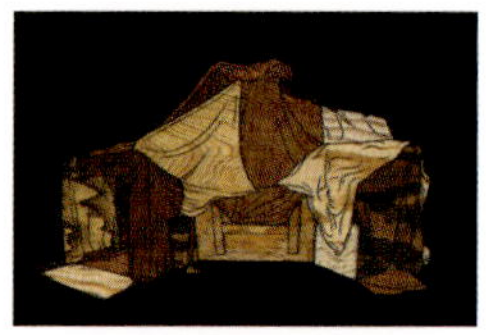

p 19 20x30,5 cm | 2010

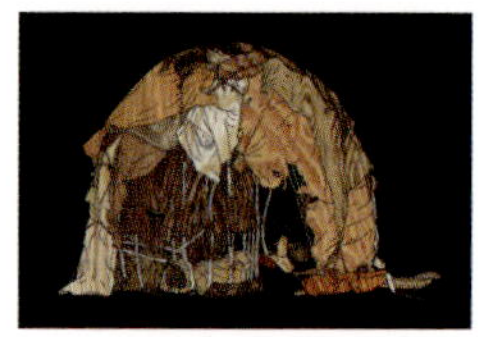

p 20/21 30x44 cm | 2013

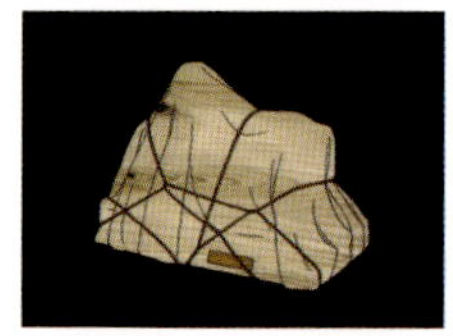

p 22 22x30 cm | 2012

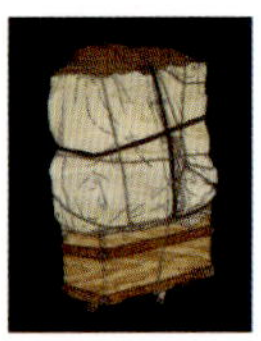

p 23 33x42 cm | 2012

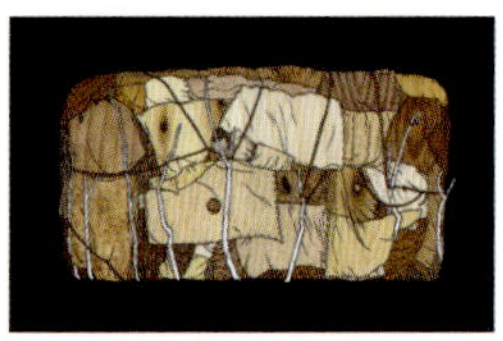

p 24/25 26x42 cm | 2012

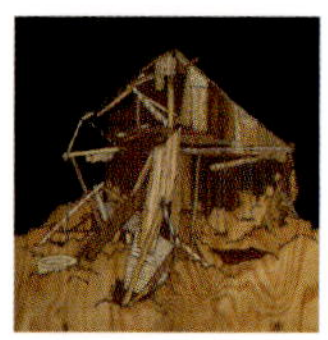

p 26 39x40 cm | 2012

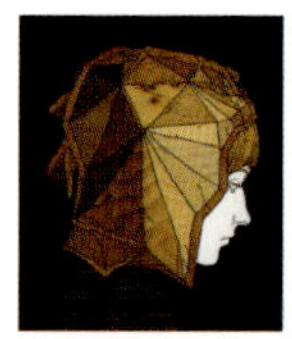

p 27 23x19,5 cm | 2012

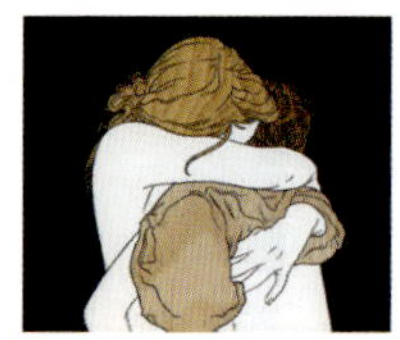

p 28 21x25 cm | 2015

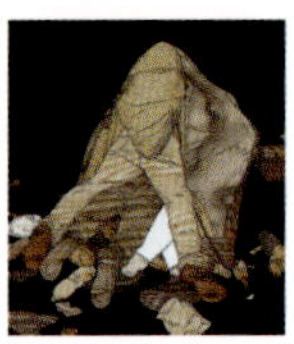

p 29 45x40 cm | 2013

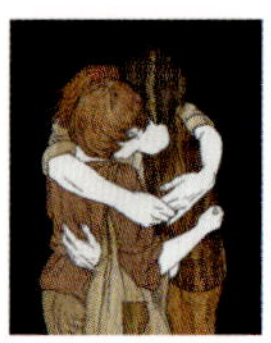

p 30 30x25 cm | 2013

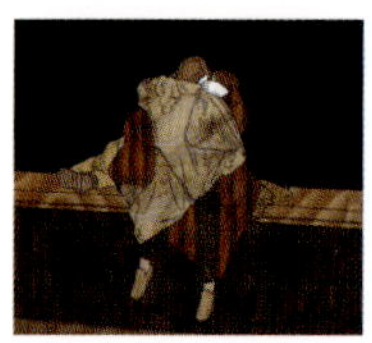

p 31 25,5x30 cm | 2013

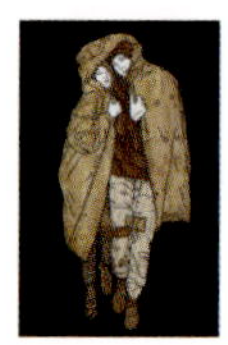

p 32 29,5x19,5 cm | 2013

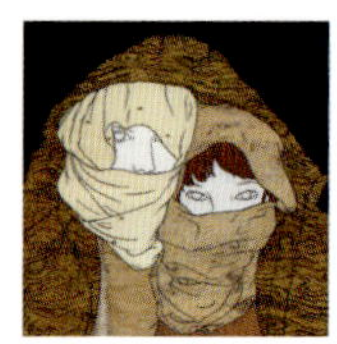

p 33 35x35 cm | 2010

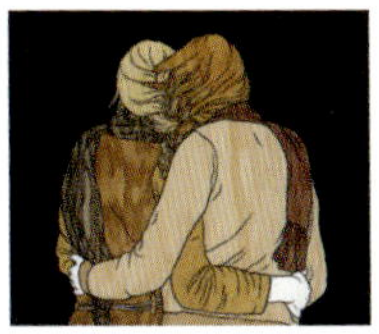

p 34 18 x 21 cm | 2015

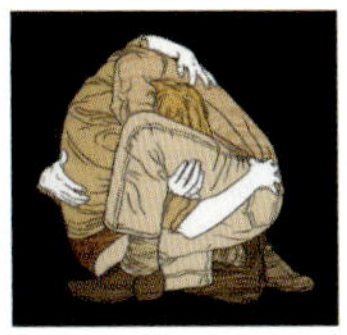

p 35 21 x 22 cm | 2015

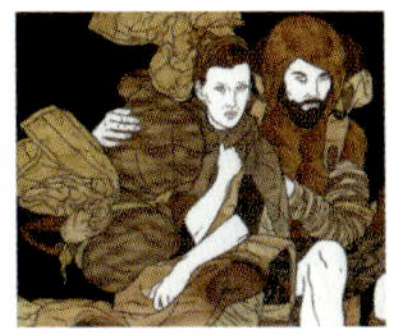

p 36 23 x 27,5 cm | 2011

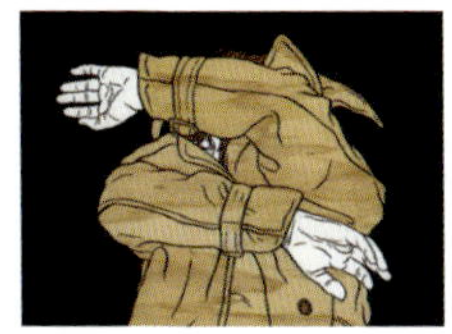

p 37 13 x 18 cm | 2014

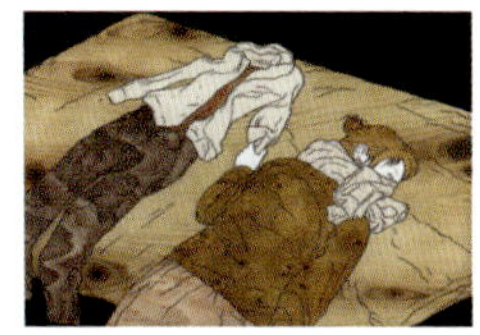

p 38 34,5 x 24 cm | 2013

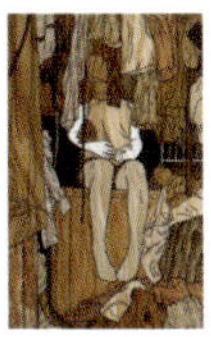

p 39 34 x 22 cm | 2013

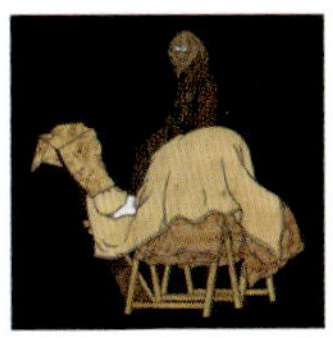

p 41 30 x 30 cm | 2010

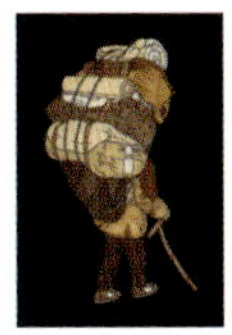

p 42 30 x 20,5 cm | 2012

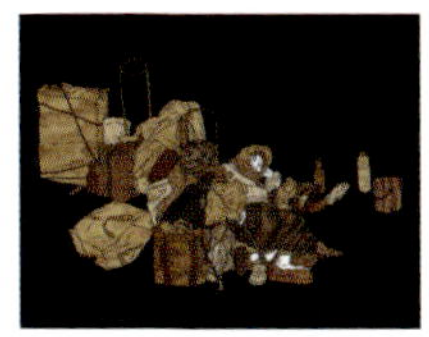

p 43 32 x 41,5 cm | 2012

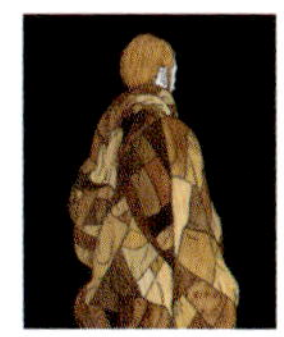

p 44 27 x 23 cm | 2012

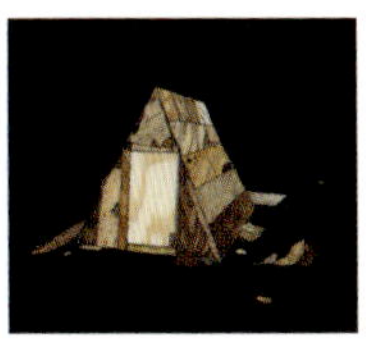

p 45 35 x 40 cm | 2012

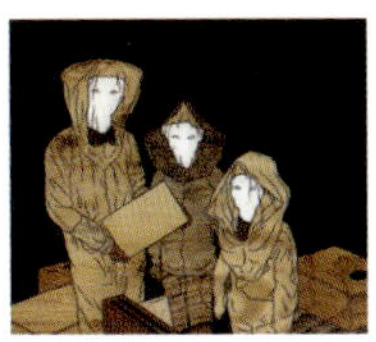

p 46 22 x 26 cm | 2011

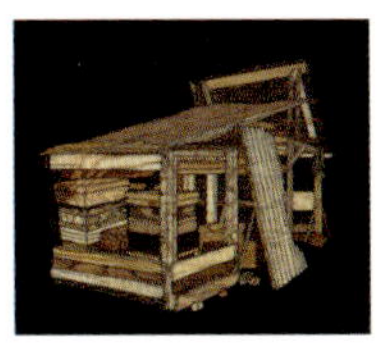

p 47 35 x 40 cm | 2010

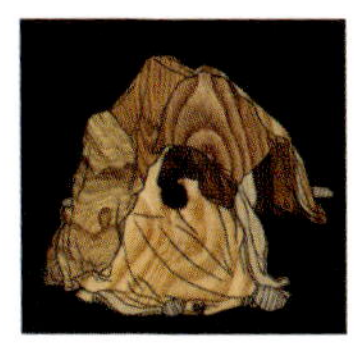

p 48 18 x 16,5 cm | 2010

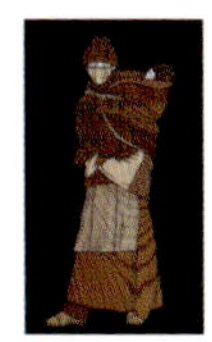

p 49 27,5 x 16 cm | 2013

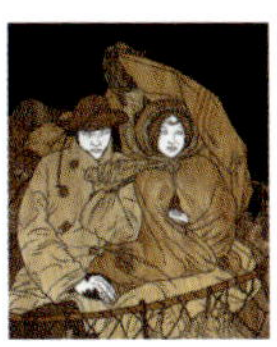

p 50 29,5 x 25 cm | 2012

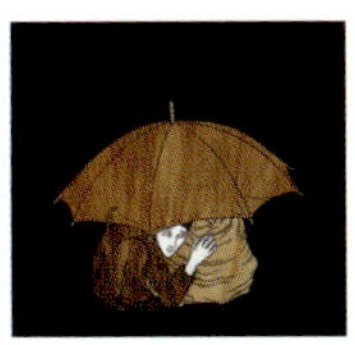

p 51 23 x 25 cm | 2012

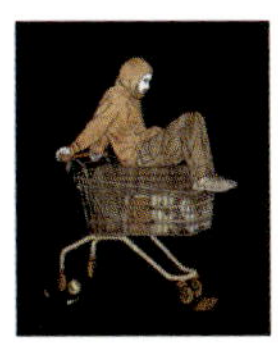

p 52 35 x 29,5 cm | 2015

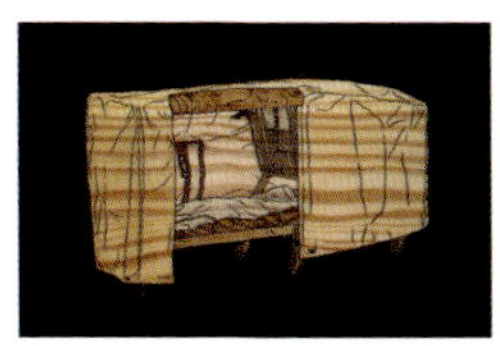

p 53 22 x 34 cm | 2013

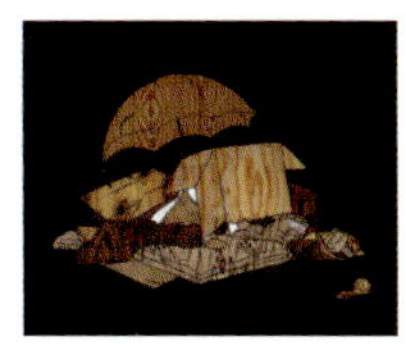

p 55 40 x 50 cm | 2010

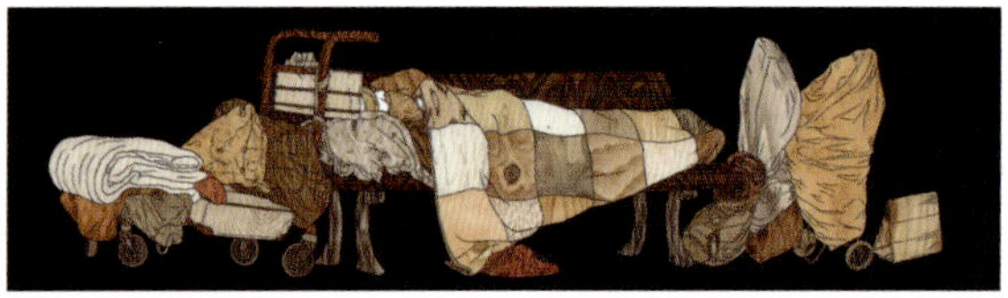

p 56/57 84 x 23 cm | 2013

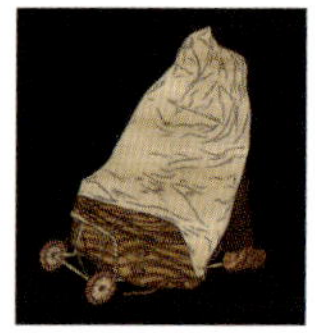

p 58 27 x 25 cm | 2013

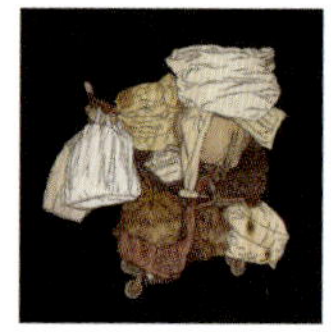

p 59 35 x 35 cm | 2013

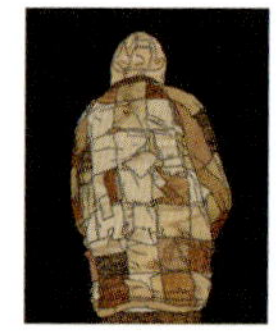

p 60 28 x 22 cm | 2015

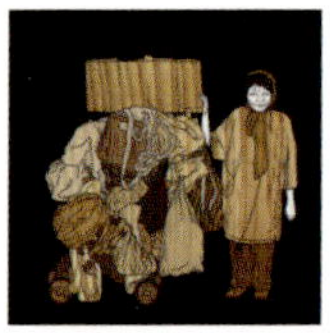

p 61 28 x 29 cm | 2010

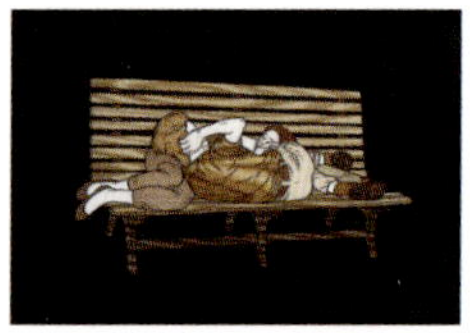

p 62 16,5 x 24 cm | 2013

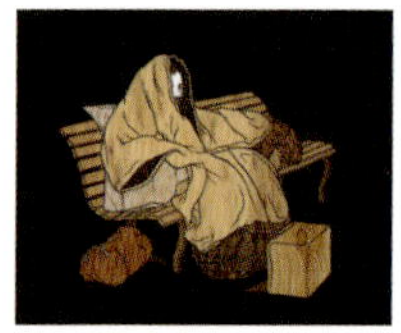

p 63 18 x 22 cm | 2014

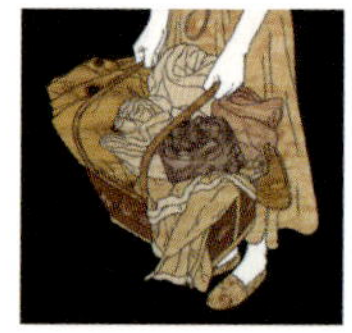

p 64 26 x 27 cm | 2015

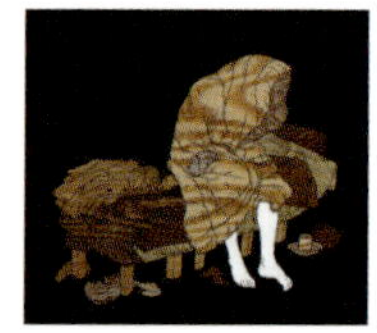

p 65 39 x 41 cm | 2010

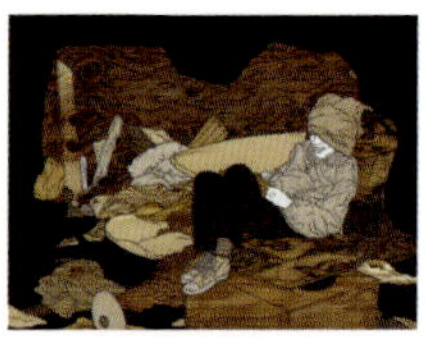

p 66/67 37 x 50 cm | 2011

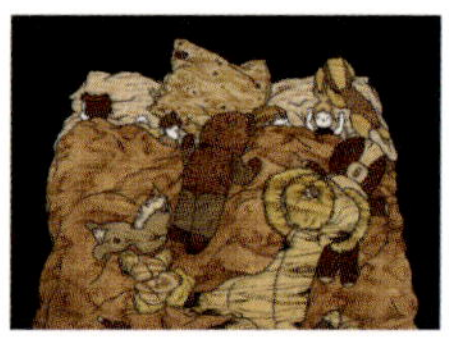

p 68 21 x 30 cm | 2010

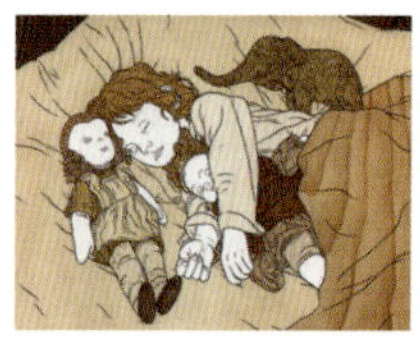

p 69 22,5 x 29,5 cm | 2013

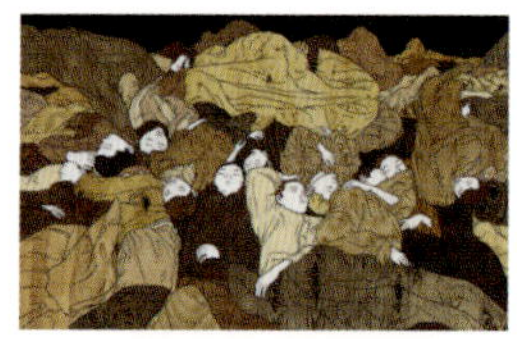

p 70 50 x 80 cm | 2010

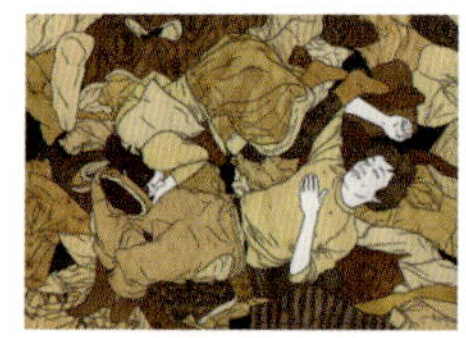

p 71 35 x 50 cm | 2010

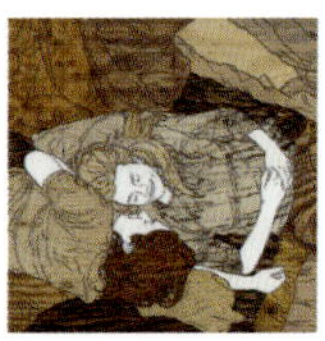

p 73 40 x 40 cm | 2013

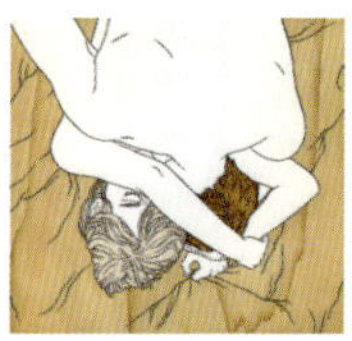

p 74 27 x 30 cm | 2013

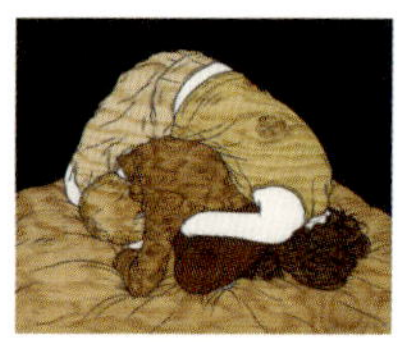

p 75 22 x 27 cm | 2011

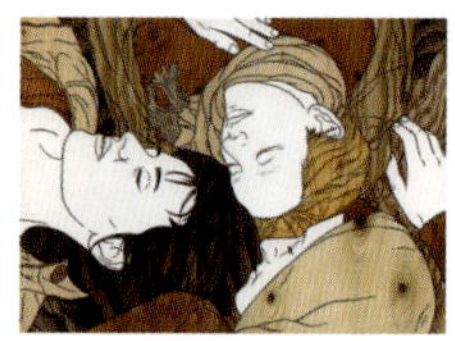

p 76 19,5 x 27 cm | 2013

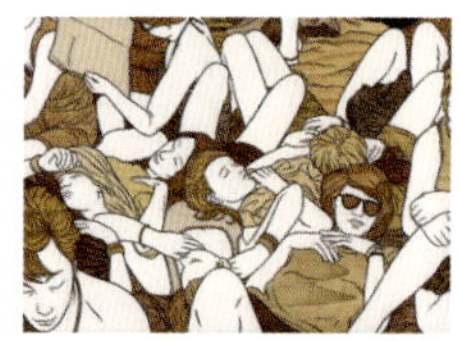

p 77 30 x 21,5 cm | 2012

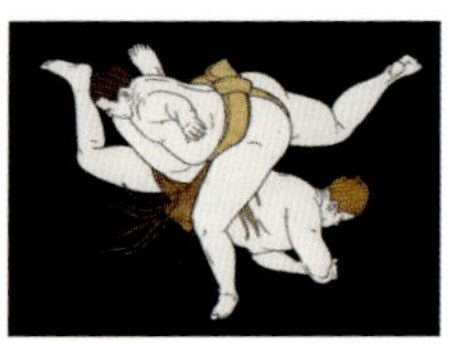

p 78 21 x 30 cm | 2011

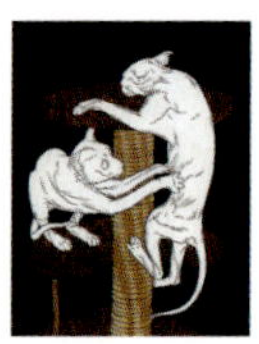

p 79 35 x 28 cm | 2010

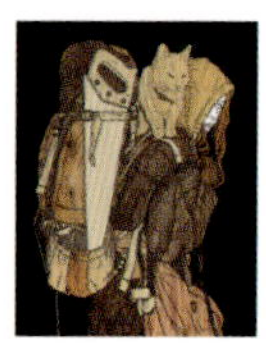

p 80 36,5 x 29,5 cm | 2015

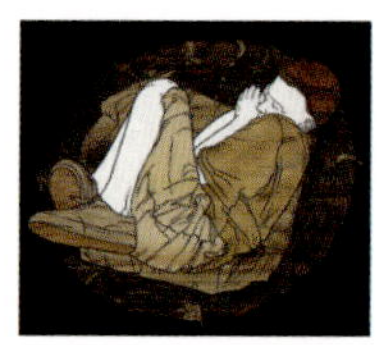

p 81 19 x 21,5 cm | 2012

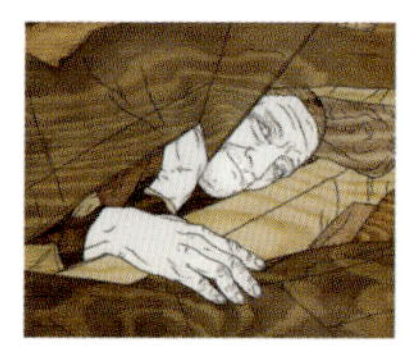

p 82 25 x 30 cm | 2012

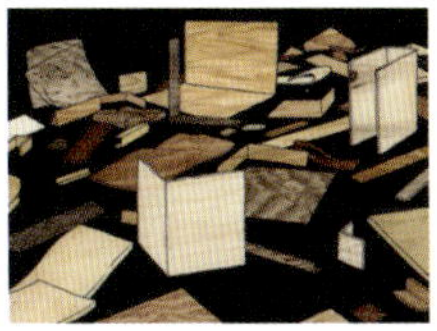

p 83 22x30 cm | 2012

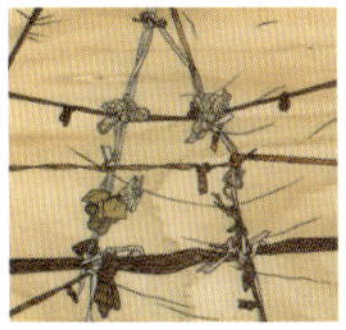

p 84 22x23,5 cm | 2015

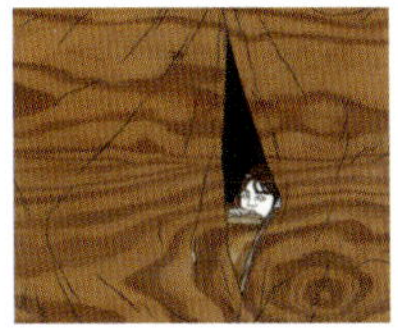

p 85 20,5x25 cm | 2013

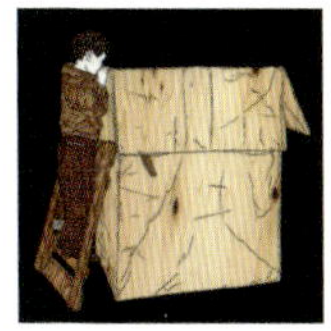

p 86 25,5x25,5 cm | 2012

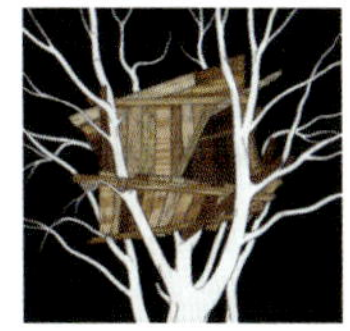

p 87 25,5x25,5 cm | 2012

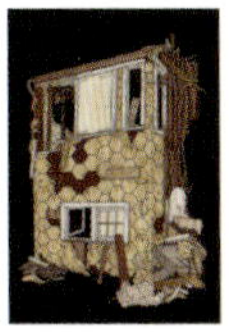

p 89 59,5x42 cm | 2012

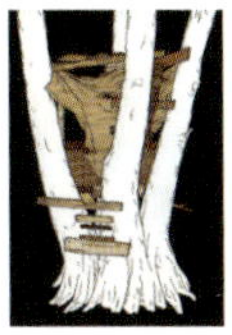

p 90 35x24 cm | 2012

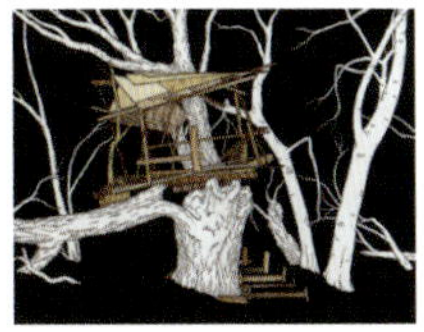

p 91 32x42 cm | 2013

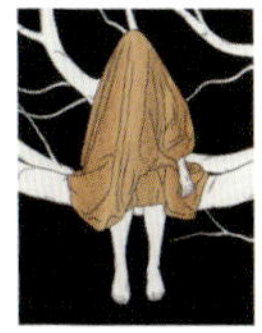

p 92 29,5x23 cm | 2013

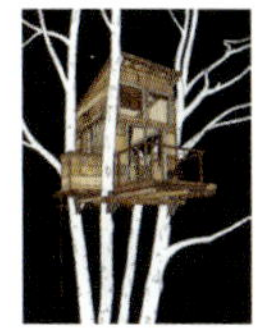

p 93 46x60 cm | 2013

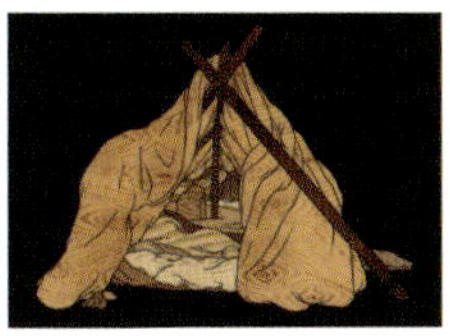

p 94 21x30 cm | 2010

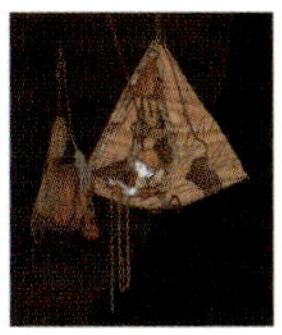

p 95 38x32 cm | 2013

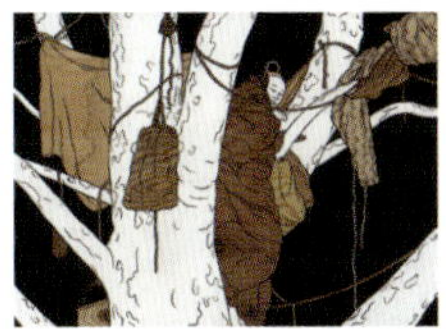

p 96 21,5x30 cm | 2014

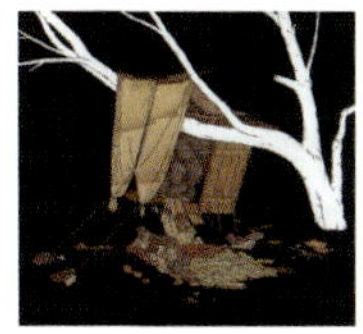

p 97 50x55 cm | 2013

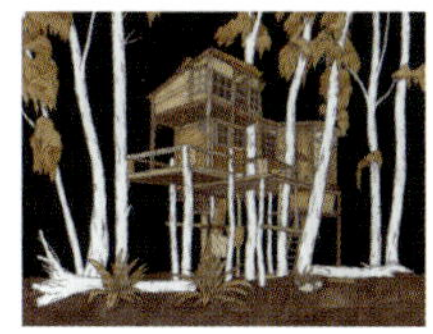

p 98/99 50x65 cm | 2013

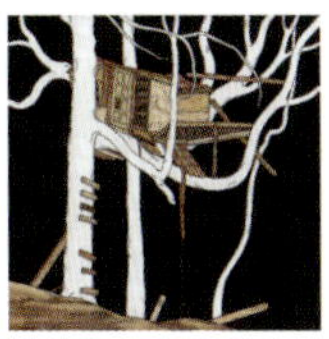

p 100 40x41 cm | 2012

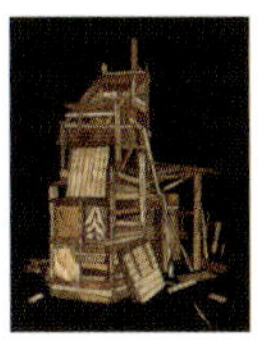

p 101 45x35 cm | 2010

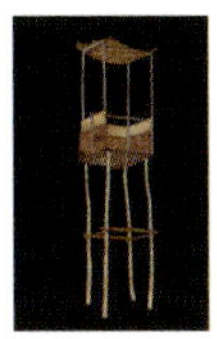

p 102 44x28 cm | 2012

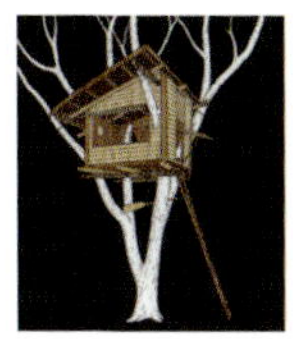

p 103 50x45 cm | 2012

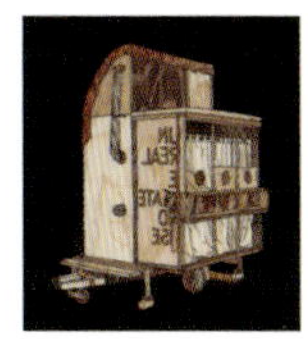

p 104 38x35 cm | 2014

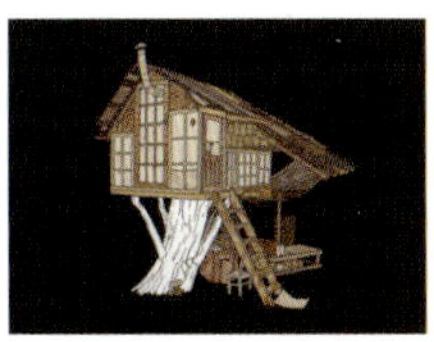

p 105 37,5x50 cm | 2014

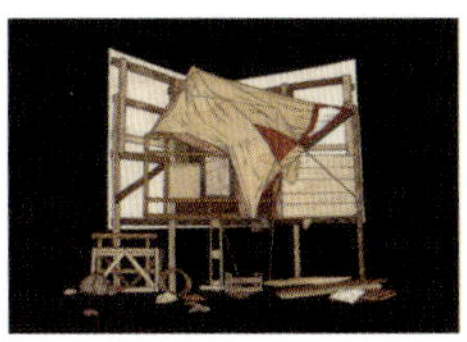

p 106/107 56x80 cm | 2013

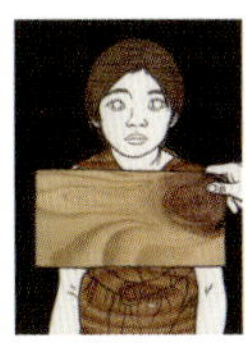

p 109 15x12 cm | 2010

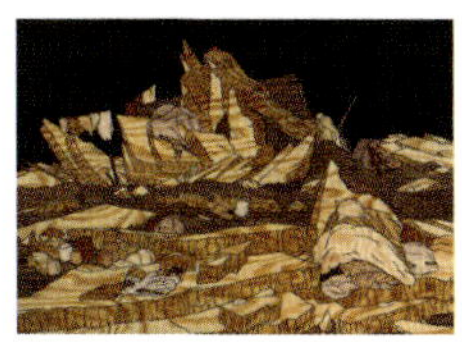

p 110 35x50 cm | 2010

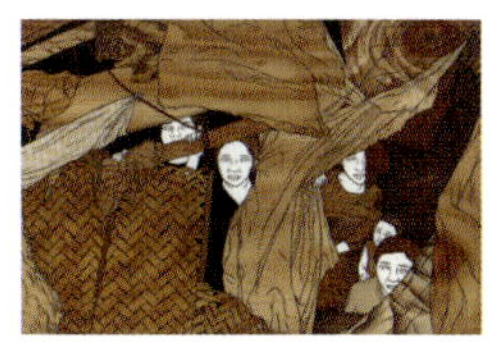

p 111 26x39 cm | 2010

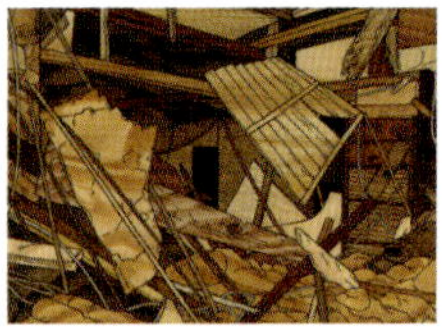

p 112 23,5 x 33 cm | 2012

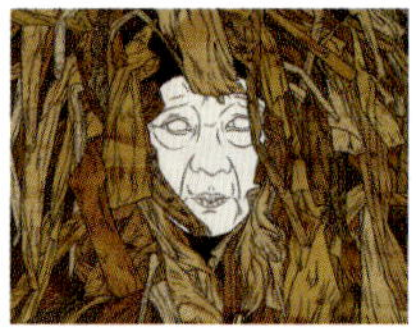

p 113 18 x 23 cm | 2012

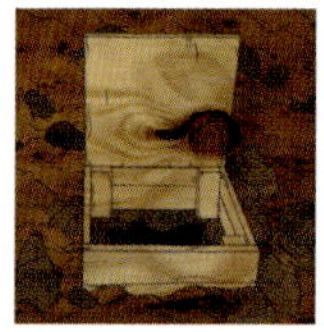

p 114 30 x 30 cm | 2010

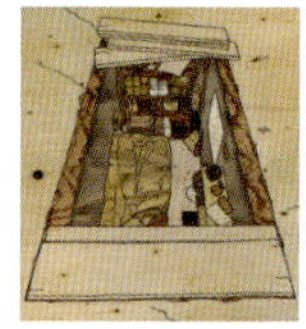

p 115 40 x 36 cm | 2014

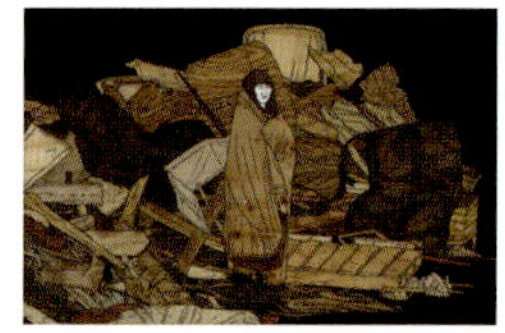

p 116/117 35 x 52,5 cm | 2012

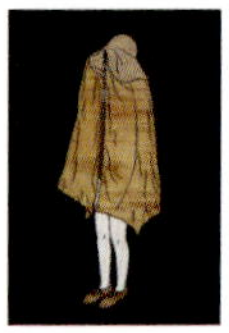

p 118 19 x 13 cm | 2012

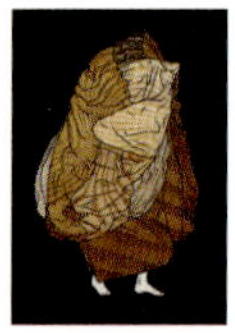

p 118 22 x 15,5 cm | 2010

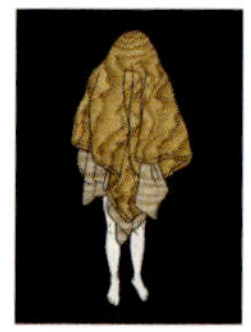

p 119 19 x 14 cm | 2010

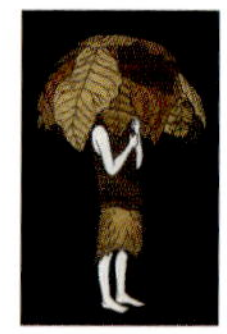

p 119 19 x 14 cm | 2012

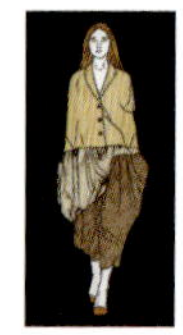

p 120 19 x 13 cm | 2012

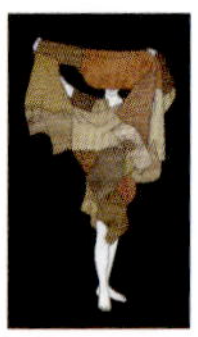

p 121 55 x 33 cm | 2013

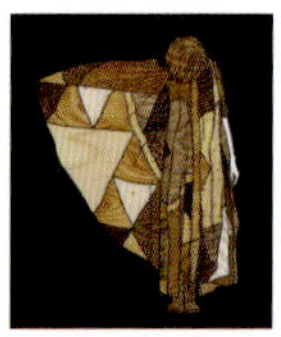

p 122 17 x 19,5 cm | 2013

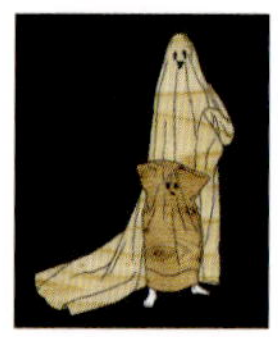

p 123 17 x 20,5 cm | 2013

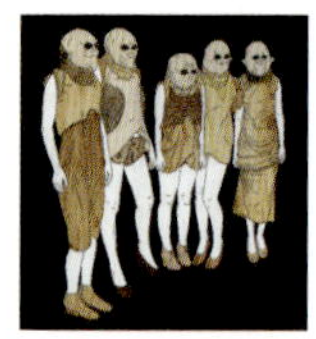

p 124 52 x 50 cm | 2013

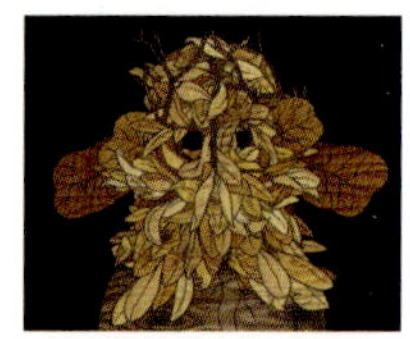

p 125 16,5 x 20 cm | 2013

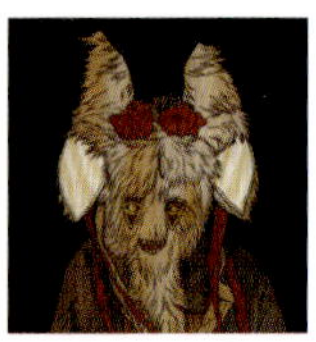

p 127 17 x 17 cm | 2013

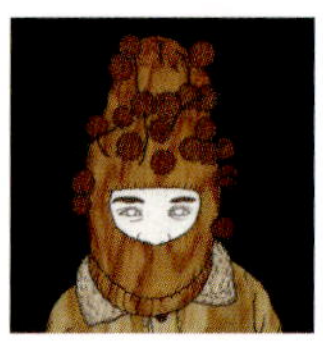

p 127 17 x 17 cm | 2013

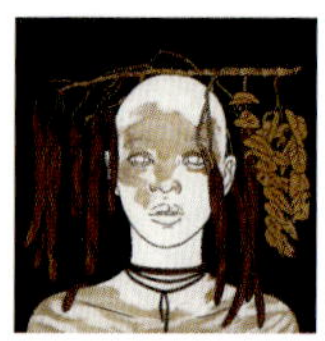

p 127 17 x 17 cm | 2013

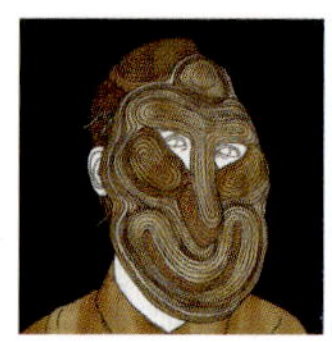

p 127 17 x 17 cm | 2013

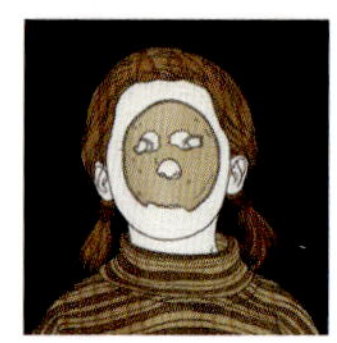

p 128 17 x 17 cm | 2013

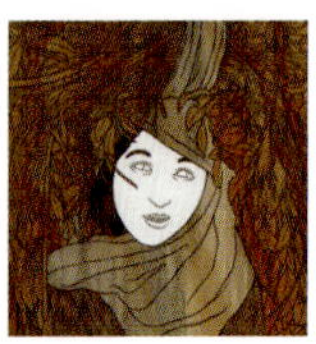

p 128 17 x 17 cm | 2013

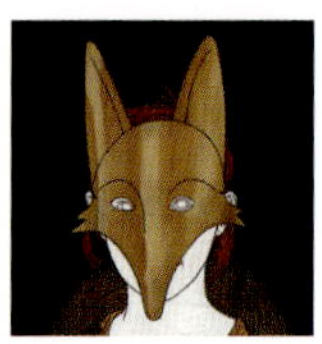

p 128 17 x 17 cm | 2013

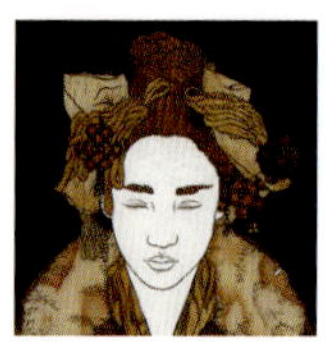

p 128 17 x 17 cm | 2013

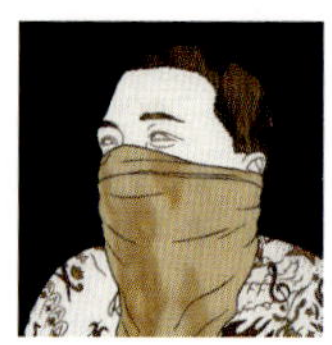

p 129 17 x 17 cm | 2013

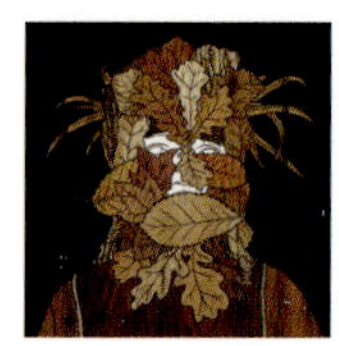

p 129 17 x 17 cm | 2013

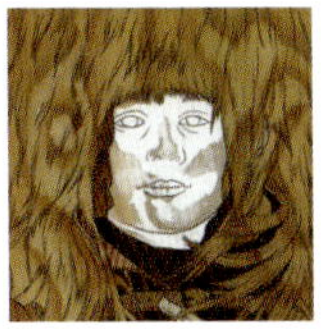

p 129 17 x 17 cm | 2013

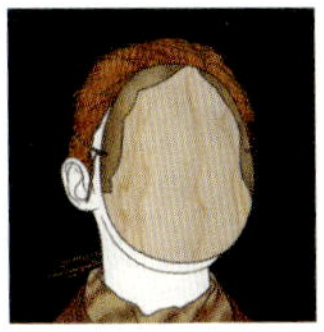

p 129 17 x 17 cm | 2013

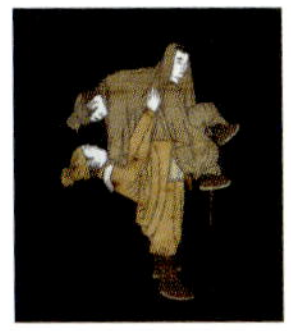

p 130 40 x 35 cm | 2013

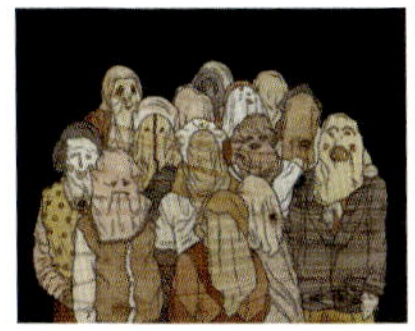

p 131 26,5 x 33 cm | 2015

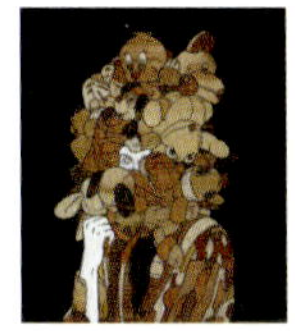

p 132 30 x 26 cm | 2015

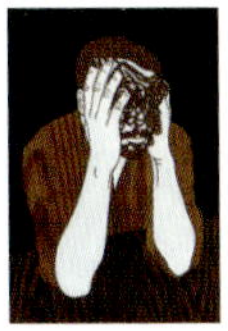

p 133 27,5 x 19 cm | 2012

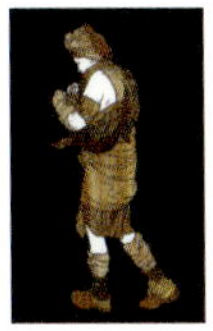

p 134 21 x 33 cm | 2011

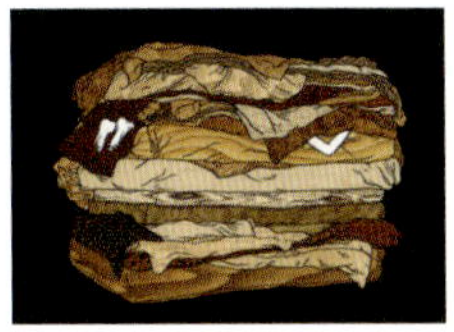

p 135 21 x 30 cm | 2010

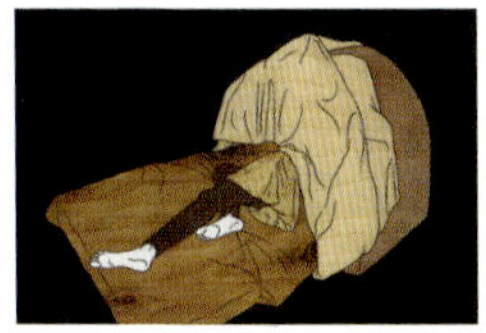

p 136 19 x 29 cm | 2012

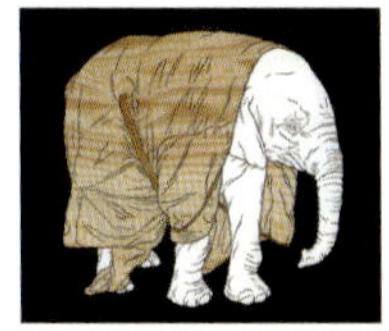

p 137 30 x 35 cm | 2013

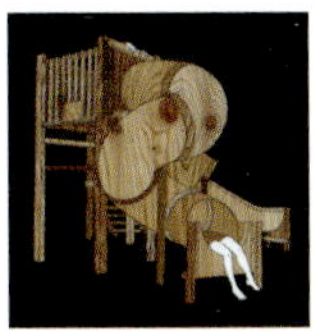

p 138 50 x 50 cm | 2013

I personally felt that a box,
far from being a dead end,
is an entrance to another world.
I don't know to where,
but an entrance to somewhere,
some other world.

from The Box Man
by Kobo Abe

Seeking Shelter

Interview with moki
conducted by Anika Heusermann

Anika Heusermann: In your current pictures one sees tent- and hut-like structures, built from wooden panels and boards, cardboard, boxes, tarps and fabric – some fragile and temporary, some sturdy with a door and a window. There are many different forms of housing and shelter. In your book you quote from Gaston Bachelard's philosophical study in several passages. Here, Bachelard describes the house as a place of intimate space: caves, nooks, a roof – spaces that have emerged from an existential necessity, as that also created by animals, for example nests or shells: *Shelter* as protection or refuge. What were you seeking in this context?

moki: In *Shelter* I have explored forms and variations of going into hiding, shielding, disguising or veiling yourself, including the apparent need to create a space of your own and to feel safe there. For me this suggests many facets and this is why I also wanted to break it up thematically. So I asked myself: What actually is a shelter? How far does it reach? The artist Hundertwasser once stated that human beings had three skins. They would be born into the first, the second would be their dress and the third the façade of their home. You could extend this idea by the social environment in which a human being moves about, meaning family and friends. In the images I create we see depictions of places and structures that offer refuge in a purely physical sense, such as huts or masks, covers or clothing, but also illustrations of a social shelter, one that could manifest itself through the comforting sensation of belonging to a group. You will also see body postures that offer a kind of protection and a place to stay, like huddling or curling up or even just closing one's eyes.

A. Heusermann: This describes quite a broad movement, from the self-chosen haven of a tree house or playful children's hiding places to images that express the necessity of creating a protective space for oneself. There are, indeed, quite a few images that point specifically to your social urban environment in Hamburg and Berlin, to people wandering about with a shopping cart filled with their belongings, people on a park bench… Was this a vital impulse for the *Shelter* series?

moki: I live in Berlin near the Oranienplatz, where refugees lived in self-constructed huts and tents for about two years. The huts were mainly built of whatever these people could find in the streets: recycled materials like plastic tarps, wooden panels, mattresses, scraps of fabric – a patchwork of different surfaces. What was being improvised by the floors and walls of these structures, was a private sphere, a safe haven. By creating a boundary, a dialectic of the inner and outer world arises, and it is this field of tension that has caught my interest. The hut-like constructions visualize the needs of the refugees and show the creativity that people can develop due to a lack of materials or being reduced to the use of found materials. Objects that are produced with recycled materials – whether it be dwellings, toys, household appliances or pieces of clothing – all are bearers of a hidden beauty and the undeniable misery behind it. It seems as though there was a longing inherent in them, like an unbreakable willpower to create something and to realize it, no matter what resources are available.

A. Heusermann: Where do you find images and motifs?

moki: I constantly find picture material that I want to keep – for the most part I do research on the Internet, but I also find it important to be out on the streets with my camera. So, before starting a series, there are always images that I can revert to and which I then assemble into something new. My *Shelter* archive is filled with very different pictorial subjects: from the covered-up sewing machine to self-made head protection, to the designer shopping cart which in a flash can be transformed into a place to sleep. Quite chaotic, one might think, but I have a guiding principle that ties it all together. Many of the motifs of *Shelter* can be traced back to real encounters with shelter seekers and refugees, some whom I approached directly and captured their portraits. On other occasions, I took photos without getting noticed, like the portrait of Napuli L. (page 96) or Karl-Heinz G. (page 56-57). In addition, there are those friends of mine who know what I am working on and who send me pictures. This is how I received a photo of a homeless woman from Kreuzberg who wraps herself in scraps of fabric that she artfully ties in knots and then plucks them off and burns them (page 134). I also stage my own settings whenever I can't find a fitting template. The entwined lovers amidst a sea of blankets (page 73), for example, and the man in the patchwork blanket (page 44) are such composed pictures. And there are paintings that have other paintings as an inspiration, such as Caspar David Friedrich's *Sea of Ice* (page 110) or *The Last of England* by Ford Madox Brown (page 50).

A. Heusermann: Does it matter to you to know when, how and where a picture originated? When you pick up on an old painting or a photo that you took yourself or that someone has sent you, you usually know much more about the context of its creation than when you find an image on the web. There, the motif can be much more detached from its origin.

moki: What counts is the image or impression that the picture conveys. Of course, the background interests me, but, for the viewer, most of this context is lost through my reworking and recomposition of the picture. The compilation of pictorial content of the most varied origins, rearranging it, reworking it and thus bringing it forward into a new overall context is an important part of my art. The juxtaposition of subjects and the resulting associations lead to a concentration capable of conveying the image and basic feeling, the underlying sentiment. In the case of *Shelter*, it is this deep longing for comfort and security that resonates in the work.

A. Heusermann: A longing for security is also reflected in the utilized material and in the visual execution of the work: Warm tones of brown and beige are predominant and the appeal of wood or wood texture is a defining characteristic. The images, in general, and the puzzle-like impression of the shelter constructions, in particular, also recall the woodcut technique. In some cases one only detects on a closer view that the wood texture is painted – even if on wood. How did this come about?

moki: In painting, there are lines that almost always get lost in the process of painting over the preparatory drawing. These lines have captured my interest. In color woodcuts they remain intact. I thus experimented and, with the support of my uncle, a restorer, initially made an inlay of wood veneer. But instead of producing other pictures using the inlay technique, I went back to drawing or rather painting and replicated the wood grain in this manner. To this effect, I paint wood onto wood, often using scraps from a carpenter's workshop or found pieces of wood. I orient myself on the grain direction – picking up on the existing veins and knotholes – or I alter them. In the past, furniture was painted to resemble fine wood and thus make it look more precious. This imitation, the art of deception or disguise, is fascinating to me. The way the puzzle-like surfaces of wood grain interlock recall camouflage patterns, thus a disguise. It is the disguise, the imitation of a surface, that creates a protective space. When disguised you become invisible, and being invisible makes you safe and invulnerable. This aspect connects my last book *How to Disappear* with *Shelter*.

A. Heusermann: In *How to Disappear* this chameleon-like blending-in with the environment was more a vanishing without a trace, a dissolving or decomposing in nature. The work dealt with conditions of being on the margins of defined identities, with ghosts as the "other" of identity. In comparison, your current pictures are more concrete and physical.

moki: It appears as though my artistic work has come a bit closer to the real world. *How to Disappear* explores strategies of escapism – the flight from the world, getting away from reality. The new concept is more tangible. Where there were figures dreaming themselves out of reality, unreal places and a host of fantastic creatures, the images in *Shelter* reflect a realistic context: victims of natural disasters, wars or poverty, but also places and situations capable of counterbalancing the worst aspects of being at the mercy of harsh circumstances. The challenges we are facing on a global scale seem so overwhelming that it is tempting to escape into the private sphere and to "disappear". It is, therefore, all the more important to recognize the forms of protection, refuge and safe havens, and not to recede into one's own shell. This recognition can be a powerful motivation for action.

Published by GINGKO PRESS

Gingko Press, Inc.
1321 Fifth Street
Berkeley, California 94710 (USA)
Phone: (510) 898-1195
www.gingkopress.com

Gingko Press Verlags GmbH
Schulterblatt 58
20357 Hamburg (Germany)
Phone: +49 (0)40-291425
www.gingkopress.com

Editorial Support: Anika Heusermann
Layout & Design: Anna Dütschler

Printed in China
First Edition

ISBN 978-1-58423-578-1

moki would like to thank
Anika Heusermann, Adam Bergholm, Anna Dütschler, Mo Cohen, David Lopes, Margaret Morton, Florian Gläßel, Amber Morris, Tony Moss & Oi Va Voi, Hiromi Nishiyama, Andrew Hosner, Lena Frohne, Guido Spriewald, Heribert Kraft, Minh An Szabó de Bucs, Tara McPherson, Ben Zhu, Ralf Krüger, Michael Thomas, Jürgen Große, Richie Löffler, Helene Boström, Charlotte Ekbom, Yoichi Osaki and my family

Thank you all for sitting for your portraits
Marco & Luise, Thomas, Kostja, Lily, Alex, Tim & Marie, Gabi, Emma, Anneli, Rita